INCIDENT COMMAND: TALES FROM THE HOT SEAT

Incident Command:
Tales from the Hot Seat

Edited by
RHONA FLIN
University of Aberdeen

KEVIN ARBUTHNOT
West Yorkshire Fire Service

Ashgate

Published by
Ashgate Publishing Limited
Gower House
Croft Road
Aldershot
Hants GU11 3HR
England

Ashgate Publishing Company
131 Main Street
Burlington, VT 05401-5600 USA

Ashgate website: http://www.ashgate.com

British Library Cataloguing in Publication Data
Incident command : tales from the hot seat
   1.Emergency management - Psychological aspects 2.Leadership
   I.Flin, Rhona II.Arbuthnot, Kevin
   363.3'4'525

Library of Congress Control Number: 2001098445

ISBN 0 7546 1341 0

Printed and bound in Great Britain by MPG Books Ltd, Bodmin, Cornwall

Contents

Acknowledgements

We would like to express our gratitude to Sandra Duffield who spent many hours reformatting chapters, resizing diagrams and generally ensuring the perfection of our finished typescript. She accomplished this with tremendous patience and good humour and without her efforts, this volume would never have been completed. Margaret Crichton assisted with proof-reading and we appreciate her careful scrutiny of the typescript. Our thanks also go to our contributing authors who in the best spirit of command completed their chapters in an efficient and timely fashion, we are grateful for their candour in sharing the best and worst moments of their command experiences. We must acknowledge the advice and assistance provided by our commissioning editor at Ashgate, John Hindley who provided unfailing encouragement from the start of this project.

Part I
The Nature of Command

1 Introduction

Kevin Arbuthnot and Rhona Flin*
West Yorkshire Fire Service
*University of Aberdeen

The Purpose of the Book

One of the most critical factors in crisis management is the skill of the incident commander. This book extends the limited literature on this subject by examining personal experiences of incident command from a range of professions and attempting to reconcile these with an academic analysis of the subject. The editors are a practising fire commander and an academic researcher who has a particular interest in the psychology of incident command. From our two different backgrounds, we had both searched with limited success for case material on incident command, which could be used for training and research purposes. In the available accounts there is some recognition that non-technical or 'soft' skills feature in effective command, but the precise role that these play is often shrouded. This results in the practice of incident command having acquired the status of an art as much as a science, with many practitioners content for this to remain the case.

There still appears to be a gap in understanding between those academics who have observed and analysed the performance and characteristics of commanders (e.g. Flin, 1996: in press), and the practitioners who train and appoint commanders, or indeed who exercise command themselves. There was some doubt about whether the careful analyses that had been undertaken in many fields of critical incident decision making, including the military and emergency services (e.g. Flin et al, 1997; Zsambok & Klein, 1997), had yet resulted in the degree of improvement and better understanding that might have been hoped for or expected. It was self evident that this was unsatisfactory, particularly if

3

training and selection processes for aspiring commanders were to become meaningful. We concluded that a single volume bringing together the academic and practical perspectives may start to bridge the gap.

'Tales from the Hot Seat' is intended to assist in advancing the application of current knowledge about the subject of command, and decision-making processes associated with it, a stage further. It presents command and decision making experiences of individuals who have held significant leadership positions in a variety of 'front-line' professions from their own, anecdotal but reflective, perspective. It focuses directly on how the decision-making challenges were dealt with at the human level.

The Focus of the Book

It is accepted that many aspects of the exercise of command can be reduced to the functional, involving systems and methods. The analytical phases of decision making associated with the strategic planning phases of emergency management are in many ways similar to the demands placed upon business leaders and other professionals. While the routine management and planning aspects of incident command are essential functions, they are different in nature to the areas under consideration in this text.

The distinguishing feature of the incident command environment is that of the rapidly developing incident where information is incomplete, stress levels are high, and there is an acceptance that not making a decision is not an option. This situation is characterised by circumstances where an individual or close knit team is called upon to prevent an undesirable event deteriorating further by getting an effective grip of the situation. It is widely believed that this requires certain social and cognitive skills as well as operational techniques, in either a single commander or a command team. We hope that this book will offer some guidance to trainee officers and aspiring commanders who are starting to build an appreciation of their realm of operation and the challenges they will face. In addition, the case studies should assist the researcher who has little first hand knowledge of the reality of command in the services. They provide the context for the observations and measurements taken from field or simulator research, and give some deeper feel for the subject matter. It will hopefully enable the researcher or student of command to stand back from the detail of decision making models and the complexity of organisation charts and procedures, in order to reflect upon the qualitative aspects of command as well.

How the Book is Structured

The book is presented in three main parts, which are described briefly below, followed by a more detailed overview of each contribution. The first part consists of three chapters written by very experienced emergency services commanders who have discussed the nature of command. These are wide ranging examinations of the role of the commander and the command organisation. Consideration is given to the significance of teams, including their training and development, how to build them and keep them together in difficult times, as well as the likelihood that they operate best as a distinct entity in critical situations. We see examples of positive and negative role models and how lessons can be drawn from both. The issue of stress in individuals and teams features widely and the subject is considered in the context of selection and training. Situation and risk assessment is introduced in the context of decision-making psychology. Contributors consider the development of a command perspective, the role of the various command levels, and finally, the critical aspect of implementation, i.e. making things happen by connecting the command process to action on the ground.

In the second part, six personal case studies are presented, written by expert commanders from a variety of disciplines. Their work domains represent very different environments, but their common focus is on the human aspects of the role of operational incident commanders. They all draw upon long careers, during which they have often been called upon to demonstrate command competence, with accounts of specific incidents or periods where they have had to exercise effective command. This is more than merely a description of the role or a catalogue of the decisions taken at a particular incident or series of events. This is a section that will give the reader a feeling for the real-world practical difficulties and pressures of discharging such roles and responsibilities. It is also hoped that it will show why the management of critical incidents, in all their real-world complexity, can be quite different to the way it is portrayed in operational manuals. The references to leadership, decision making, teamwork, stress, communications, procedures, pre-planning are part of the backdrop rather than the main focus of the study. It is intended to offer the reader a clear feel for how experienced practitioners do what they do.

The third part of the book reflects in some detail on the issues raised by the commanders and applies existing scientific knowledge to the development of commanders. These chapters consider the extant research on decision making and how this might be trained, techniques to optimise command team training derived from empirical investigations, and the use of the latest command simulators designed to enhance commanders' cognitive and social skills. In the final chapter, the editors review the lessons learnt from the previous contributions with a view to the selection, training and development for the commanders of the future, as well as the legal, media and political pressures they will be expected to face.

The Nature of Command

This section of the book sets out to define the issues that are recognisable as being relevant to operational incident command in its many guises. *Arbuthnot* surveys a range of topics starting by considering the language used in discussing command and control and exploring the definitions that are commonly used. Any study of command cannot be conducted without considering the parallel topics of leadership and management, and the question is raised about whether these are distinct areas or more closely interwoven than some academic works might imply. It is well accepted that effective command is linked to decision making; Arbuthnot develops this into a discussion about what the nature of the critical decision really is and argues that it often revolves around some form of risk assessment in complex situations and dynamic timeframes. *Sarna* draws on his experiences as a senior police commander and offers a carefully considered analysis of situation assessment and 'command perspective', as well as the correct focus of training and exercising. *Brunacini* writes in his inimitable and incisive way presenting some timeless truths about the exercise of command and gives several pointers about getting to the vital stage of the process: making things happen.

Case Studies of Command

The generic nature of incident command has not been fully explored. From a psychological viewpoint, commanders need to be able to demonstrate very similar kinds of thinking skills and interpersonal skills. Humans respond to demanding situations in broadly similar ways, whatever style of uniform they are wearing. Leaders have been shown to make gross errors of judgement in civilian, military and industrial emergencies (Flin, 1996). But very limited learning is shared across organisations that are required to manage crises. So one of the objectives of this volume was to distil the common elements of the incident commander's role across professions. To that end, we invited six eminent commanders to share their personal experiences of 'sitting in the hot seat'. We asked them not only to describe one or more episodes they had taken command of, but also to reflect on how they felt while this was taking place. We were particularly interested in their leadership style, how they made their decisions and the manner in which they worked with their teams. Their colourful accounts of how they managed some very difficult situations serve to demonstrate their ability to critically examine their own leadership actions – the hallmark of an expert commander. At the end of each chapter, they were each asked to offer some specific advice gleaned from their own experiences at the sharp end.

In the first case study chapter, *Moore* graphically describes his experience as a Metropolitan Police Commander tasked with the management of an escalating public order problem, namely a riot on the streets of Notting Hill in London. This is followed by another emergency services commander, *Davies* who provides a lucid account of one memorable night as the senior Fire Commander at one of Britain's largest industrial fires at the *Associated Octel* plant in Cheshire. In both cases, the commanders highlight the factors that led to a successful conclusion of the incidents and their own subsequent reflection on what contributed to this outcome.

We were more than a little conscious that while there may be a general absence of literature on emergency services or other commanders, the same cannot be said of their military counterparts. However, with the exception of McCann and Pigeau's (2000) enlightening new volume on modern military command (see chapter 14), most military autobiographies by commanders focus more on the technical and political aspects of their command career rather than the psychological dimensions. Moreover they have generally not considered how their acquired wisdom might transfer to other professions. We therefore asked two distinguished military commanders, who had experience of training industrial managers for emergency command roles, to contribute to our book. *Larken* has crafted a most insightful chapter outlining his experience as a Royal Navy Captain during Britain's conflict with Argentina over the Falkland Islands. *Keeling* reviews two demanding periods of command during his career with the Royal Marines. He shares striking recollections of command in Northern Ireland and of a subsequent episode in Northern Iraq which clearly highlights the enormous difficulties for modern military commanders in charge of peace-keeping operations. In both chapters the commanders reveal the significant demands of very lengthy operations, when they are on duty for weeks or months, in contrast to the operations lasting for hours or days, which typify public and industrial emergencies.

The two final case studies serve to illustrate that critical incident management is not just the province of the military and emergency services commanders. A surprising number of professionals work in remote or secure locations where they will be required to manage an emergency, even though this is not the main focus of their occupation. *Lodge* shares his experiences as the commander of a commercial aircraft, laterally as the Captain of a *British Airways* Boeing 747. He gives a very frank account of the challenges which pilots are trained to deal with as they operate their aircraft across the globe. He describes a domain where commanders must be sufficiently competent to deal with any unexpected situation with the crew and equipment they have onboard, and where there is no opportunity to summon additional resources or to hand over to a more senior commander. Finally *Coyle* provides a perspective from a commander in an even more private workplace – a maximum security prison. Drawing on his experience as a Prison Governor, he graphically portrays a series of major incidents he managed in the role of on-scene incident commander. We had also hoped to include an account of emergency management from a site manager of a hazardous industrial site, but unfortunately none of the companies approached were willing to provide this.

Developing Commanders and their Teams

In the final section, *Crichton* and *Flin* review the latest research into naturalistic decision making and underline the importance of situation assessment skills for command decision making. Having considered the relative merits of different methods of reaching a decision, they suggest how commanders' decision skills may be trained. Incident commanders never work alone and *Salas, Cannon-Bowers* and *Weaver* present a state of the art review of their extensive research into high performing teams. As specialists in team training, they offer a number of tried and tested methods for enhancing command team performance. In another training application, *Crego* and *Harris* return to the subject of decision making and show how command skills can be effectively developed using the latest evolution of a computer-based command simulator.

Finally, we review the main lessons, which we extracted from all our contributors' chapters and synthesise these against a framework of organisational issues and command skills. We also look to the future and consider the political, media and legal challenges that are beginning to frame the incident commander's task. These issues invariably arise post-incident, where the actions and decisions of the commander are reviewed in the cold light of hindsight. The commander's credentials, in terms of qualifications and experience, are beginning to be routinely challenged, along with a new scrutiny of decision making, team work and leadership skills. Experience shows that our society, as represented by the judicial process, is becoming less tolerant of the failures that can arise in the 'heat of battle'. The incident commander's environment is one in which the enlightened approach of management gurus such as Peters and Austin (1986, p. 180) to 'celebrate failure' is not an option, because in the realm of the military, the emergency services and the high reliability industries, that failure is likely to cost lives and cause suffering. The need for robust, defensible procedures to select, train and assess the competence of incident commanders has never been more apparent. We hope this volume makes a contribution to this important endeavour.

References

Flin, R. (1996) *Sitting in the Hot Seat. Leaders and Teams for Critical Incident Management.* Chichester: Wiley.

Flin, R. (in press) Piper Alpha. Decision making and crisis management. In A. Boin, U. Rosenthal, L. Comfort (Eds.) *Managing Crises. Threats, Dilemmas, Opportunities.* Illinois: C.C. Thomas.

Flin, R., Salas, E., Strub, M. & Martin, L. (1997) (Eds.) *Decision Making Under Stress.* Aldershot: Ashgate.

McCann, C. & Pigeau, R. (2000) (Eds.) *The Human in Command.* New York: Kluwer/ Plenum.

Peters, T. & Austin, N. (1986) *A Passion for Excellence. The Leadership Difference.* London: Fontana/Collins.

Zsambok, C. & Klein, G. (1997) (Eds.) *Naturalistic Decision Making.* Mahwah, NJ: LEA.

2 Key Issues in Incident Command

Kevin Arbuthnot
Deputy Chief Fire Officer, West Yorkshire Fire Service

Kevin Arbuthnot started his career as a firefighter in North Yorkshire in 1975 and served in mainly operational roles in Tyne and Wear, Warwickshire, West Midland. He was promoted in 1994 to Assistant Chief Fire Officer, head of operations, in West Yorkshire. Kevin was promoted to Deputy Chief Fire Officer, again head of operations, in West Yorkshire in January 2000.

West Yorkshire Fire Service is a busy urban fire brigade serving a population of 2.1 million. The head of operations is responsible for 50 fire stations, which are staffed by 1680 firefighters and officers.

Kevin Arbuthnot gained a Master of Philosophy degree from Sheffield Hallam University following research into the application of the principles of empowerment into a fire brigade command structure. This work came to fruition when the West Yorkshire Incident Command System, which he developed in the brigade from 1994 onwards, was accepted as the basis of the Home Office's system published in the Fire Service Manual on incident command. He was selected for and completed the Brigade Command Course in 1994, is an active member of the Institution of Fire Engineers, which elected him to a Fellowship in 1999, and represents the UK's Chief and Assistant Chief Fire Officers Association (CACFOA) on incident command issues. In 2002 he was awarded the Queen's Fire Service Medal.

www.wyorks-fire.org.uk/ics

Introduction

Before the contributors to this volume present their experiences and thoughts in the following chapters, it may be useful to survey some of the issues surrounding the topic of incident command and the decision making of commanders. Although much has been spoken and written about the burdens and responsibility of command, it is probably worth stating the main areas that concern commanders as well as trainers, supervisors and psychologists. This chapter will, therefore, give some attention to the issues of language and terminology used to define and discuss the subject, try to establish what the differences in 'command levels' actually mean, and consider the importance of leadership in the process of command. It will examine the area of risk assessment, which is at the heart of the command task in most of its guises and, finally, consider how a commander may demonstrate competence to discharge the responsibilities that have been vested in him or her.

To demonstrate the universal relevance of the key principles of command, most of the material which will be presented in the following chapters has been drawn from experienced commanders in a wide variety of disciplines and services. However the author has drawn on many lessons learned in the research and implementation phases of the West Yorkshire Incident Command System, the forerunner of the incident command model for fire brigades' operations featured in the Home Office's Fire Service Manual (1999). The kind of decision making that is under examination is principally exercised at the tactical and operational (or task) levels. It would probably be helpful to offer some distinction between these levels. The skills of task, or operational, level command tend to be demonstrated within smaller groups of people, and the nature of the task more immediately obvious to the commander. The requisite skills are more likely to be linked to those of competence to perform the task in hand, e.g., captaining the aeroplane, capping the oil well or dealing with one sector of a large fire.

At a tactical level, however, the challenge, and requisite skills reflect the fact that the commander often has to operate in a more physically remote fashion, balancing strategic imperatives with front line realities in a constrained time frame, utilising as much of the limited information that it is possible to get about circumstances on the ground. These skills are identifiably managerial in nature, but are still characterised more by dynamic timeframes and environments rather than static ones. Studies by Klein Associates in the USA have estimated that for fireground commanders, the situation they faced changed on average five times per incident (Klein, 1998, p. 6). This was further confirmed by Burke and Hendry (1997)[1] in the UK who reported finding that during 205 'incident minutes', over seven recalled incidents, there were 121 significant events, or one per minute, 45 seconds. Gilchrist (2000) in the UK identified a 'transaction rate' greater than one event per minute (p. 142, 287). It is in such dynamic environments that the skills, experience, training and other characteristics associated with the effective commander come into play and either succeed in bringing the situation to a satisfactory conclusion, or otherwise.

Some would argue that this state of affairs is not exclusive to the emergency services. An example of an environment that bears many similarities may be the trading desks of the stock exchange, where very rapid analyses, with often very high financial stakes involved, are conducted every hour of every day (Lindley, 1999). However, this book confines itself to the context of the emergency incident or military operation, where the consequences of ineffective command can result in tangible and immediately visible losses, injuries and sometimes deaths, that are played out against political backdrops and which usually have legal consequences. Much of the important research that has been conducted into command decision making has recognised this essential characteristic of the commander's operating environment (Klein, 1998, p. 6). In this respect the question of society's acceptance of risk arises. How the public, in the form of the media and the courts, perceive risk and commanders' responses to it is often the key determining factor in any judgement about whether a commander was successful or not. The consequences of actions taken in the areas under consideration are commonly held to be in a different category to similar losses in the business world, which, being financial in nature, may be larger but of lesser social impact.

Terminology

The subject of command and critical incident decision making is to some extent complicated by the lack of a standard terminology for the topic across all the professions and disciplines where it is exercised. Therefore, before we consider these issues, it is necessary to establish some common language and definitions that will be helpful in the course of reading this book. The sections that follow provide working definitions for (a) command and control, and (b) command organisation.

Command and Control

The term 'command and control' is often used to describe a single set of processes, whereas it can be argued that the functions are separate and distinct. In the UK, the term 'control' is usually associated with the direction of discrete parts of an operation. For example, in firefighting, the management of breathing apparatus teams towards a desired operational objective within defined procedures is considered to be a 'control' issue as an element within the overall command of the incident. Attempts have been made to standardise the use of these terms, notably in the Home Office (1997) publication 'Dealing with Disaster'. This includes a glossary (pp. 15–17) that includes the terms 'command' and 'control'. It defines 'Command' as:

> The authority for an agency to direct the actions of its own resources (both personnel and equipment).

'Control' is given as:

> The authority to direct strategic and tactical operations in order to complete an assigned function and includes the ability to direct the activities of other agencies engaged in the completion of that function......

Carl von Clausewitz (1976, p. 121) suggested that command is:

> The authority vested in an individual for the direction, co-ordination and control of resources.

Other interpretations do exist, however, few fire brigades' operational orders contain any attempt at a distinction between the two terms. The current usage of the words 'command', 'control', 'leadership', etc. is sufficiently complex that a dictionary definition alone will not serve to remove confusion, and possible contradiction, altogether. The issue has been further complicated in recent times by references in business school and management gurus' publications, usually critically, about 'command and control' organisations. Robert Heller (1994, p. 60) stated, for example, that the ideal of Total Quality Management was a:

> shift away from command and control (or order and obey).

He added his belief that:

> Many managers can't make the move.

Such observations are usually references to a management style that prevails in some organisations, rather than to the tasks they undertake or functions they perform. The term 'command' is, more often than not, in such cases, used in a pejorative way, frequently suggesting a link to a hierarchical, inflexible 'control' style of management. This can no doubt be supported by countless examples of managers or others in authority having used a 'don't do as I do, do as I say' approach. It is equally likely that it reflects the fact that military or emergency service operations have to be controlled in an apparently autocratic style, although practitioners in those disciplines would differ from that simplistic viewpoint.

The difference probably originates in the wide variety of ways in which the terms are applied. It is a generally accepted fact that officers in organisations such as the military, fire service and police have certain demands placed upon their leadership and capacity for sound decision making which are different to the demands placed upon their otherwise equally skilled and sometimes equally hard pressed commercial counterparts. The environments that military and emergency service commanders operate in often demand that the directions of the incident commander, whose role is to have an overview of the entire incident or field of operations, be followed immediately and without question, in a traditional 'command and control' style. It is also fairly widely accepted that in such types of operation, an adequate decision made and executed quickly will often satisfy a need more effectively than one delayed by an attempt to find an ideal approach, or to conduct consultation.

Again, this situation is not exclusive to the military, fire or police; managers in high-risk process chemical companies, offshore oil installations and many other hazardous operations also have to operate in this environment. Business schools' management case studies are, however, rarely based on such examples. This particular set of considerations constitutes part of the rationale for the use of the term 'command organisation' in this book. It helps to differentiate it from the context the term is used in when it is applied to traditionally hierarchical organisations, not necessarily being of a uniformed or disciplined nature or engaged in hazardous work.

'Command Organisation'

Some further exploration of the nature of a command organisation and the terminology involved in this is, therefore, called for. These organisations are probably most easily exemplified by the uniformed emergency services and the military. One of the obvious characteristics of a command organisation is a clear command, and usually hierarchical, organisational structure, which is different from a command style. The latter can be found in any number of organisations, not traditionally or necessarily associated with the exercise of command. A command structure reflects certain legitimate operational considerations of an organisation. These may include the need for narrower spans of control to facilitate effective communications, a dynamic and challenging operating environment, and the need to be able to adapt and vary procedures quickly and sometimes, of necessity, without consultation with strategic managerial levels. The structure also reflects the thread of accountability, which runs in a strong and unbroken way throughout the entire organisation.

In its wartime or operational mode, the military is assumed to be a definitive example of the command style and structure. The fire service, like the police and any other emergency service are also a clear example of an organisation that, historically, has discharged its statutory functions by exhibiting the traits of a command organisation. Without these traits, the integrity of the operation, and the service provided to the public may be compromised, as might the health and safety of the operational crews. The need for the maintenance of a command structure need not preclude elements of devolution and empowerment. However, using the case of the fire service as an example, this demands that certain boundaries be drawn to the process, and constraints identified before an initiative is launched. Sanders summarised the situation as follows:[2]

> Discipline versus empowerment is not a contradiction, but a contrast; not only the function but also the symbolism is essential to a command organisation.

In this volume, both Keeling and Larken reinforce the point and make reference to the fact that the exercise of command and the role of orders in the military is quite different to the way it is perceived in civilian life. Once the parameters have been clearly drawn, there is considerable freedom to operate within them.

Levels of Command

In the UK's approach to integrated emergency management, it is common to identify three levels of command, i.e., strategic, tactical and operational (or task) levels. Figure 2.1 below represents the levels of command as defined in the UK Home Office's (1997) publication 'Dealing with Disaster'. In broad terms, those at the front line are operating at the task or operational level. The role of the tactical level of command is to co-ordinate the actions of the operational commanders. The tactical commander's goal is to implement the plans and achieve the objectives set by the strategic level of command. The strategic level is the most senior in the organisation and may involve political and policy level decisions that extend beyond a single organisation. How this broadly accepted division and labelling is adopted and adapted by the various agencies that use it depends on many things, including the organisation's size, its sphere of operations, its nationality and the nature of the task.

Figure 2.1: The levels of command

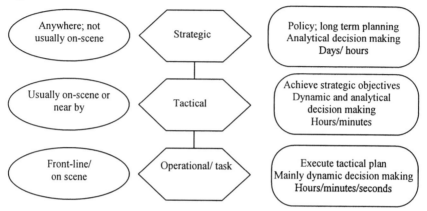

Following a spate of civil disturbances UK-wide in the 1980's, the police forces in England and Wales introduced a protocol reflecting the three main levels of command. This was simply that strategic level was described as *Gold Command*; tactical level as *Silver Command*; operational level as *Bronze Command*. The terms were probably borrowed from Plato's political and philosophical work 'Republic' in which he outlined his 'Magnificent Myth'.[3] The myth consisted of a suggestion that God had added gold to the composition of those destined to be rulers, the Guardians, silver into the veins of the soldiers and executives, the Auxiliaries and bronze into the veins of the farmers and workers. Although Plato's model is clear and understandable, it is probably a little elitist for the contemporary palate.

Gold Command: Strategic As a result of this police-led initiative, some of the emergency services in England and Wales now also use these labels for the three main levels of command. In this type of command protocol 'Gold', or strategic command, is invariably exercised at a distance from the scene of the incident. It often includes representation from municipal management, normally the chief executive of the city or district, and from other stakeholders in the event, such as water companies, government departments, the Environment Agency and the like. It is intended to take the longer view of the situation; the time frame of Gold, or strategic command, is in days rather than hours or minutes. The task is to anticipate and project developments. Based on reports and other data from the incident, the task of the strategic commander is to plan and progress towards the restoration of normality. Decision making is, of course, involved, but it is, on the whole, analytical, managerial skills that are brought into play rather than the dynamic decision making skills of the on-scene or tactical commander. Life and death decisions, therefore, are not very often made at Gold command level, although such dilemmas may from time to time arise.

Silver Command: Tactical The tactical commander co-ordinating the areas of operational activity on the ground is known as 'Silver'. This is the co-ordinator and supporter of the sectors where operational level command is being exercised. Tactical commanders are more routinely faced with critical decisions; also operations in sectors need to be co-ordinated. Decisions made by sector commanders have to be appraised and validated and, if they are not satisfactory, amended. A tactical commander has to operate on the basis of information received and be prepared to make an intervention if necessary. Consequently, the time-scales that come into play can swing suddenly from a managerial or analytical pace to a dynamic one. For example, where an intervention is necessary it often has to be made in seconds and minutes or it is likely to be too late to be of value. Often, for the same kind of reasons, the opportunity to discuss or consult is limited or non-existent.

Bronze Command: Operational The direct supervision and co-ordination of operational crews working in sectors is the responsibility of the operational or 'Bronze' level command, where the focus is at the team and task level. They are nearer the sharp end of whatever action is underway and they direct the teams performing the tasks. Bronze commanders must make immediate decisions and dynamic risk assessments. They must motivate and control crews doing difficult or dangerous, and sometimes distressing, work. They frequently have to lead from the front. Consequently, their timeframes are routinely short, with quick decisions and ongoing appraisals of the developing risks being necessary most of the time.

The Appropriate Location for Commanders at each Level. Although at first consideration this may appear to be an issue of detail, the location of each commander in a hierarchy such as the Gold, Silver, Bronze model is significant and can influence the success or failure of the mission. In the context of UK emergency service operations, at Gold or strategic level, as described above, there is rarely a problem with commanders physically departing from the command headquarters, wherever that is, to intervene on scene. From time to time there may be the need to accompany a VIP or undertake a visit for the purposes of supporting morale of the team, perhaps during a protracted operation at a train or air crash. The temptation to become involved in the tactical decisions, or worse, operational decisions, is reduced, as, despite sophisticated telecommunications, the commander does not have a direct 'feel' for the status or tempo of the on-scene operation.

At Silver, or tactical, command, however, both the temptation and the opportunity exists for the commander to interact directly with the Bronze, or operational, commanders. This may be dressed up as 'leading from the front' or establishing a command presence, and may be difficult to discern from genuine attempts to achieve those worthy objectives. Nevertheless, research on many fire incident grounds in West Yorkshire has shown that incident commanders (tactical) have a tendency to wish to gain their own appreciation of the situation in the highest risk or activity sector (their perception thereof). They then tend to interact at that point until they feel the urgency or activity level has diminished. Such excursions frustrate the line of command and their absence from the tactical command post results in the incident commander being denied the overview of the incident as it develops, and the command team being denied the commander's presence and input to the dynamic situation (Gilchrist, 2000, p. 58). This phenomenon is discussed later in this chapter in another guise; that of the commander's 'comfort zone'.

The Significance of the Three-Tier Command Model

This schema satisfactorily embraces the theory of the inter-relationship between the levels of command. However, the impact made by the strategic level can depend to a large extent on the nature of the tactical operation. There are two common patterns.

The first type is characterised by the situation of the planned operation where the strategic level of command operates exactly as its place in the command hierarchy would suggest. Strategic command would develop a plan with a clear objective, allocate resources and support the tactical commander in the operations being conducted to execute that plan. It is notable that given modern communications, there are many examples today of strategic command attempting to direct or otherwise interfere with operations in progress (see Coyle and Larken in this volume) and hence the boundaries as described become less distinct.

A second pattern is where the event being managed was not planned and the command structure builds from the bottom, or front-line, upwards in a reactive way. This is the normal experience in fire service operations, but is not uncommon for the police or, presumably, the military as well. In such scenarios, the strategic command has to start by embracing what the tactical commander has put in place. Attempts to directly influence operations at too early a stage, if at all, can be difficult and sometimes ill advised, due to the availability and quality of information, as well as the commander's lack of proximity and incomplete appreciation of conditions on the ground.

The challenge of command does not necessarily become easier or more difficult with changes in level, but it definitely changes in nature. This is one of the reasons that experience of, or competence in, command at tactical or task/operational levels does not prepare a commander, by itself, for the exercise of strategic command, or indeed vice versa. It is of limited value to study command in one of these dimensions if the focus should really be on another. Although trainee or junior officers may learn much about leadership and gain inspiration from generals and chiefs, if they want to learn their trade as task level commanders, they must study the experiences and methods of commanders at that level. Naturally, competence at any level is enhanced by a full appreciation of the skills and roles of commanders immediately above and below one's own position in the structure.

It is important that the levels of command and the labelling associated with them are respected but not embraced over rigidly. There are many cases where the roles overlap out of necessity. Clearly, if a decision of a tactical nature has to be taken when only an officer of junior rank is in attendance, it will be taken. Equally, someone operating at strategic, or more probably tactical, levels will not overlook an urgent issue that arises that can easily be dealt with without further ado, merely on the grounds that it is not of the appropriate level of importance. This flexibility must not, however, be allowed to mask a common command failing of becoming too involved at lower levels for too long for no other reason than the individual is experienced and comfortable making decisions at that level. This will be discussed later when considering the 'comfort zone'. For a variety of reasons, on the whole political rather than practical, there has been an apparent desire to move away from the use of the terms 'Gold, Silver and Bronze'. This is evidenced partly by the absence of these terms in the latest editions of the government's publication 'Dealing with Disaster' (Home Office, 1997). However, as is usually the case, if something works, it sticks, and such has been the case with this terminology. Practitioners, in the larger police and fire services in England at least, have found it to be a useful form of shorthand to describe the banding of command responsibility and the interrelationship between the levels. The issues associated with this will be further developed later in the chapter.

It should be borne in mind that different models and terms are used by different agencies across the world. Later in this volume contributors will use language and labels that are different to those outlined in this three-tier model. For example, Brunacini describes a system of fireground command that uses terms such as 'strategic' for the role of the on-scene incident commander. This is clearly different to the UK model and, understandably, bears closer comparison with US military practice, but it works well for them. As long as everyone likely to be involved is using a commonly understood set of terms and reference points there will not be a problem. Clearly, it would not do for organisations that border each other or work together to have different labels or terminology.

It is quite probable that the confusion caused by the different use of the terms 'strategic, tactical and operational' is one of the main reasons that the labels 'Gold, Silver, Bronze' endure.

Having considered the structures which enable commanders to operate, it is now necessary to briefly consider what makes an effective commander, and how the quality of leadership, however difficult that is to define, features in the success of an operation.

Command and Leadership

This section explores some of the issues surrounding command and leadership. Recognising that there is a wealth of literature on these subjects readily available it is brief, but moves on to consider how these concepts are linked to command decision making, and experience of command situations, which will both be covered by other contributors to this volume in more detail.

In examining the tales in the following chapters from commanders who have found themselves in the 'hot seat', it is worth pondering what the role of a commander is. Functionally, the role is well defined. Writers such as Adair (1968) have identified possible common traits of 'leaders'; what extra or different characteristics are required to be an effective commander, and do they differ from one command task to another?

The nature of the command task itself differs significantly from organisation to organisation, and, in the context of this volume, more significantly from managerial level to managerial level. Command is often considered to be one and the same thing as leadership, itself a notion that is difficult to define. Command also encompasses a large part of what is considered to be management, whereas again the converse is not necessarily the case.   However, effectiveness in command of critical incidents is indisputably closely linked to the ability, under pressure, to assess the situation, and then make, and implement, effective decisions. The scope for confusion is understandable, because the exercise of tactical and operational command embraces many of the elements associated with the concept of leadership. Command is arguably distinct from leadership, just as it is similar to, but distinct from, management. There have been many attempts to reduce all of these things to lists of functions. To do that is, however, misleading and can erroneously lead to the view that the exercise of command is merely the completion of a series of ticks in boxes.

Brian Fuller, the former chief fire officer of the West Midlands Fire Service and former commandant of the Fire Service College related an anecdote supplied to him when he was a student on a senior command course at the Fires Service Staff College by an old naval commander. It explored the relationship between command and leadership. The retired officer, an S.W. Roskill, said that command could be defined as 'the imposition of one person's will upon another to achieve a desired aim'. He continued by stating that if the supplementary phrase 'with the willing co-operation of the subordinate' were added, then one had defined leadership as well. Whether one agrees with that definition or not, it somehow paints a clearer and more substantial picture of the complexities involved in the issues surrounding command than the more contemporary offering from the Home Office, which offers an entirely functional picture of the role. It is likely that anyone who has had the privilege of exercising command will relate more readily to the definition offered by Roskill than that provided by the Home Office.[4]

Fuller himself adds to the picture with the observation that 'Command is about the imposition of understanding, respect and consequently, trust and not because of fear.' Some might argue that at this point Fuller has fallen into the trap of confusing command and leadership, immediately after having offered a distinction between the two. However, it is more likely to be the case that this betrays the close links between the characteristics of the effective leader and the qualitative competencies of the effective commander.

Many of the issues of judgement associated with the exercise of command require the exercise of good leadership. For example, the common dilemma surrounding the decision about whether, and to what extent in any given set of circumstances, it is appropriate to make an intervention of any kind that involves relieving a subordinate of command is something that regularly exercises senior officers. Nevertheless, experienced chiefs are in no doubt that it has to be done. Brunacini (1985) holds the view that if a commander is unwilling to change a decision of a subordinate, he should 'stay at home and watch the incident on TV'.[5] Fuller expresses the same sentiment from a different angle, stating 'it is a failure of command to accept rank and then not exercise it'.[6] It is little surprise, therefore, that the contributors to this volume, particularly those from military backgrounds, have made much play on the theme of leadership in the openings of their respective command experiences in the following chapters. Team building, morale building, developing mutual respect and understanding, all fall easily and squarely in the realm of leadership. The general debate concerning perceptions of a distinction between 'command' as a directive and authoritarian form of management, and the more positive image associated with 'leadership', whilst undoubtedly real to many, is only loosely related to the issues under examination in this chapter. Authoritarian and sometimes brutal management was fairly standard and widespread during the industrial revolution, in the factories as well as on the battlefields. Social change, particularly in the developed world has, arguably, tended to result in a convergence between the traits associated with good leadership and those associated with effective command. In short, commanders do not and cannot expect blind obedience from their followers. This is neither a problem nor a recent change. A modern commander would no doubt feel exposed leading a team of subordinates who behaved like automata. Larken develops this theme in this volume, drawing on analysis of naval command at the battle of Jutland (Gordon, 1996). Readers of Gordon's work will see that debate about the nature of command is not a recent phenomenon. In the 19th century Vice-Admiral Sir George Tryon worked hard to develop the understanding of his officers that detailed plans would not work in the 'fog of battle'. He encouraged subordinate commanders to use initiative and in extreme cases even disobey an order if circumstances and their judgement compelled them to do so (Gordon, 1996, pp. 193–214). While his views were considered radical by many in the Admiralty and the establishment at the time (Gordon, 1996, p. 262, 274), his teaching would appear contemporary to present day leaders, in the emergency services at least.

It would appear that the result of the suggested convergence of styles is that the moral authority to command resides much more firmly in the domain of the commander's skills, knowledge and experience, summarised by the term competence, than ever before. Displaying positive traits of an effective leader (as listed in many works on leadership, e.g., Adair, 1968) the modern commander must manage as leader of a consenting team in much the same way as his or her contemporaries in other fields of management endeavour. This appears then to be set as the backdrop for the subsequent exercise of operational command, with a strong implication that, without that preparation, the process would be weakened or compromised. It will be for the reader to judge whether, or to what extent, this is the case.

Students of command can successfully model decision-making processes and gauge the effectiveness of communication and information assimilation by commanders. They can construct protocols and frameworks to guide new or inexperienced commanders in the discharge of their responsibilities. However, the best that can be expected to emerge from this is a prescriptive template for future actions. The debate on the role of leadership adds a further dimension to these technical considerations, and returns to the view that the exercise of command is as much art as science.

However complex the role of the operational commander, the nature of the commander him or herself, as a thinking, feeling human, is another factor in the equation. Characteristics of courage, resoluteness and hardiness are essential features (Buck, 1999, pp.346). McCann and Pigeau (2000) go as far as stating that 'person-versus-self conflict is the single biggest factor in determining a military mission's success or failure' (pp. 3-4). Symptoms of person-versus-self-conflict include poor discipline, low morale, and a weak ethos. They surface at times of adversity. Commanders from the dynamic environments represented in this volume are accustomed to dealing with such conflict and, whereas they may not raise it in those terms, it is another facet of the human backdrop to their particular 'tale from the hot seat' that the reader would do well to bear in mind.

The effective commander's function may, to some extent, be analogous to the role of a conductor of an orchestra. Without a conductor, no doubt, music would be played from the score sheets by any group of competent musicians and the result would probably be recognisable and of some quality. With an experienced conductor present, however, it is likely that the music would be sweeter and more harmonious, the end results altogether more professional. This may be the same with effective incident command.

Decision-making by Commanders

Critical incident decision-making lies at the heart of the subject under consideration. When identifying command failures, it is invariably a flawed decision by a commander, or a failure of the commander to make a decision under stress that is at issue, or at least, the most obvious 'cause' of failure when blame is being apportioned. It is the main area where scientists have done a lot of high quality work and reached some clear conclusions, but relatively little of it has been assimilated into the training and selection of commanders. This section offers the perspective of a practising incident commander rather than an academic, but serves to demonstrate common ground in the conclusions reached. (See Crichton and Flin in part two of this volume for the academics' perspective).

Some practising commanders may question themselves about whether, in the effective discharge of their roles, they have really made decisions. They are likely to have, and done so continuously, but not necessarily in a considered and structured fashion (Klein, 1998).[7] Even when merely observing the progress of an incident that is neither demanding nor unusual, experienced commanders will be forming their own mental models of the incident. This will be automatic and will be used as the framework against which current tactics will be evaluated. It is a constantly developing and evolving set of beliefs upon which commanders will base their decisions, for example, about whether, at what point, or to what extent to intervene. More experienced commanders will, of course, benefit from their knowledge and experience in the formulation of their set of beliefs, which their subordinates will be less well equipped to do.

It is worth taking a moment to contemplate the meaning of the word 'experience' here, as it is a widely used informal measure in assessing a commander's competence, but just as widely misunderstood. Accept for a moment that experience amounts to the sum total of one's training, exposure to events, applications of one's skills, opportunities to reflect on successes and failures and the lessons that have been learned along the way. Clearly then, what it is not, is a simple function of time served, or exposure to a narrow range of events, but that is how it is often considered in common usage. Time served reflects something, but it is perhaps more in the order of a tribal scar than a qualification. The degree of correlation between a commander's competence and time served is not great.

It is not uncommon for the commander's image, or mental model, of the incident to be flawed, perhaps because the information underpinning it has been incomplete or inaccurate. This naturally leads to an insufficient or inaccurate impression of the progress of the operation that is underway. Equally, the evaluation of the risk assessment and other projections will be compromised. As well as emphasising the importance of current, reliable information, this highlights the simple fact that the commander is usually part of a team, and the whole team must function effectively if the outcome is to be satisfactory. The contributions of the experienced commanders in this volume highlight this consistently as much as anything else. The importance of the role of the team is highlighted again and again. The commander can only add value to the process if every member of the team understands the role of all the key players and the overall mission brief.

Risk Assessment

In considering the actual detail of the decision making that exercises a commander's mind at critical times, it is likely that the majority of it will involve one form or another of risk assessment. This amounts to little more than a balancing, on the basis of information which is available and able to be readily assimilated, of cost versus benefit.

However sophisticated the requirement or the assessor, the process is one of identifying hazards, and for each one calculating the severity of the consequences of it against the likelihood of it occurring, or the frequency. Consequences can range in severity from a minor inconvenience to the person performing a task, all the way to an international disaster with massive life loss. Frequencies can be labelled in ways that reflect that something occurs, say, once in ten million occasions that the task is performed, i.e., rare, all the way down to a recognition that the identified consequence might be expected every time the task is undertaken, i.e., very frequent. Clearly, the skill comes in being able to accurately identify the hazard or hazards, which involves some degree of familiarity with the processes and the environments in question, together with detailed knowledge of, or the ability to research, the likelihood of the worst case scenarios identified coming about. In this way, quantified assessments can be produced and qualified judgements then made to support a whole range of decisions about investment, deployment of personnel, protective measures and equipment, or the value of undertaking a proposed task or process.

Risk assessment is done at many levels. Children learn about it holding their mother's hand at the roadside whilst being taught about safe crossing. Risk assessment at this level is direct and personal; it often involves issues of one's own health or survival and is a form invariably associated with the use of tools, hazardous techniques (crossing the road) and processes (driving a car). It is done in a real time frame where the speed of situation development is too great for conscious analysis of the risk. If one waited at the roadside until a calculation about the relative speed of approaching vehicles from one or more directions was performed and the results analysed for a risk factor, one would wait a long time to cross indeed. The techniques of this kind of risk assessment are not the focus of this study, but have a sufficient bearing on the issue to warrant their mention. In the context of critical incident decision making this is frequently linked with actions to be taken or ordered. If there is time, it can be done using conventional analytical processes, such as decision trees, but the defining situation is one where there is not time for the use of such analytical approaches. Skilled and experienced commanders may well be able to use analysis more effectively and in more difficult circumstances than novice commanders. However, it is when there is not time for a conventional analysis of the options, either due to lack of time, scant information or other pressures and uncertainties, that the skill of the incident commander comes into play. Once again, it is a key distinguishing factor when trying to differentiate between the skills associated with command and those of management.

A dilemma that increasingly confronts commanders is to what extent it is prudent to venture into the realm of risk taking at all. In the civil emergency services in particular, health and safety law, as well as a change in what society in the guise of the courts is prepared to tolerate, has focused the minds of commanders considerably. It is not unreasonable to believe that they will feel encouraged to consider defensive tactics more frequently, and sooner, than before. This trend is further evidenced in the UK, with a corporate homicide bill being promoted as a private member's bill in parliament.[8] Judgements made in cases such as the one that followed the Digital fire in Basingstoke in 1990[9] do little to encourage bold command decision making. In that case the judgement was that the incident commander's decision was reckless rather than bold. However, the point that emerged was that to have done nothing, i.e. remain passive or engage in defensive firefighting, would not have incurred a liability, but to have made a tactical intervention which went badly wrong did result in liability. In that particular case the decision of the incident commander resulted in the fire authority incurring damages of £18 million. However, the counter view is that such challenges introduce a discipline to the fire commander's decision making process and deter complacency.

Risk assessment at a tactical or 'silver' level of command is a set of skills closely linked to situation assessment. This is a form of 'inferencing' drawing on whatever data is available and being able to make a judgement about the adequacy and accuracy of the data to make a decision that will steer the tactics of an operation, possibly exposing staff to risk in the course of the operation. This ability to take, or direct, action on the basis of limited and incomplete information, at times when there is considerable pressure and one's teams are in dangerous situations, characterises the role of the commander.

Often, although it may appear that a decision has to be taken immediately, this may not be the case. In many cases there is some degree of latitude and the experienced incident commander will exploit this to the full. It can be used to confirm data or draw in better expertise. The skilled incident commander will be able to identify when this is the case and when it is not. It is part of the commander's competence to be able to do this, as if things do not go as planned, criticism will be just as harsh about a decision taken in unnecessary haste as about a decision that failed due to it being late. Gordon Graham (2000) of Graham Research Consultants is of the view that commanders are most exposed in cases where there is high risk, little experience of the event, and also no discretionary thinking time. He makes the point that during such tense times, commanders can fail to fully utilise thinking or consultation time that may be available, leaving themselves open to greater criticism in the aftermath for not having based the decision they made more soundly on available information and expertise. He recognises, however, that all too often there is no such discretionary time at all.

Competence in Incident Command

There is an increasing demand for managers to demonstrate their competence. Often in the UK, this is against frameworks such as the Management Charter Initiative, National Vocational Qualifications and various professional institutions' continuing professional development criteria. Arguably, certain distinctions are becoming clearer. A distinction has to be drawn between the leader's overall responsibility for an operation, whether military, emergency service or commercially focused, and the competence to manage the operation, at each key level. That is the realm of the commander, who has the clear duty to demonstrate technical competence at the appropriate level, notwithstanding that both of these roles may reside in one person.

If effective command is closely linked with abilities in the area of decision making and situation appraisal, success as a commander should be discernible to the lay observer at least as much as effectiveness as a manager or leader. This is because there will usually be an immediate task or challenge that has been handled well or less well, which leads to the possibility that the whole issue can be mapped out and new commanders' performances measured against these demands. Efforts are being made to identify and map the competencies of commanders in various organisations. Later this book explores the training of senior police and fire officers in techniques of incident command and critical incident decision making, which of course requires a framework against which to build the syllabus.

In the UK Fire Services efforts have been made to map out the roles of firefighters and officers in the full range of operational and managerial tasks they perform. At the operational level this is relatively straightforward. For example, the key stages in donning and starting up breathing apparatus can be taught in a systematic way, and observed clearly by assessors as they are performed. Similarly, operating a pumping appliance, raising a ladder or performing mouth to mouth resuscitation can all be satisfactorily assessed by observation and questioning. As the tasks being assessed move away from the realm of observable tasks to issues that involve thought and decision making, the job of the assessor becomes increasingly difficult. Even outcomes are not a clear guide to the competence of the officer performing the task, as a wider range of issues obscure the thoughts, decisions and actions of the officer being assessed.

An aspect of the development of officers' competence in operational incident command that is proving complex is the ability to identify the traps that can open up for the Incident Commander. These include disorders of command including Brehmer's (1996) 'pathologies of decision making'. These include Encystment (where the commander becomes unhealthily focused on a single aspect of the operation to the detriment of others and the whole operation), Thematic Vagabonding (where the commander 'dabbles' in many issues but fails to grip the incident or make any meaningful contribution) and Refusal (where the commander will not be drawn into making a commitment to any course of action, fails to generate a 'presence' and ends up being by-passed), but also the more basic error of defaulting to operating at a level within the individual's 'comfort zone'. This is one of the more common kinds of command failure. Commanders invariably have more experience of discharging operational command responsibilities at more junior and middle levels than the senior roles they hold, as the frequency of their exposure at larger, more serious incidents is naturally less than at their previous command levels. In those roles they spent many years doing things which they became familiar and comfortable with, and built up a raft of experience and knowledge as a result. Consequently, it is not at all uncommon to see fire incident commanders wanting to direct the extinguishing operations rather than conducting the tempo of the overall operation. That is where they feel more comfortable and sure of their position, in contrast to the more rarefied climate of the command unit or other remote command post. In a study conducted in West Yorkshire Fire Service during the development of their incident command system, a researcher observed that:

> In 11 of the 18 incidents recorded the command approach adopted was that of the 'default small incident'. Observation suggests that for at least some of these incidents … 'the style was inappropriate and may have had a significant impact on the incident outcome'. (Gilchrist, 2000, p. 138, 174).

It is possibly this kind of disorder, where officers default to decision making and risk assessment in their 'comfort zone' that, more frequently than anything else, is responsible for large incidents being commanded as if they were several small ones, rather than being managed as a whole. In fireground situations the failure is more often than not unspectacular, with more junior ranks effectively bringing the incident to a conclusion, however, this is misleading in the extreme and is definitely different to suggesting that it is satisfactory. A critical situation, such as a collapse or a crew reported missing, arising in such a poorly commanded situation would without doubt be less effectively dealt with than one where the tactical commander had full control. Similar situations can no doubt be identified in other services.

Another aspect of this phenomenon is that it is interesting to observe what happens when competences start to be defined and performance measured against them. Against such frameworks, more senior officers are better able to recognise that they are no longer as current in techniques at task or operational level as they thought they were. Techniques and knowledge at their organisation's lower command levels have usually changed considerably since they left them behind, sometimes years ago. Those concerned with developing higher managerial and command competence invariably encounter the difficulties surrounding the effect of non-observable processes and even characteristics of the candidate. These include considerations such as personality traits, or 'soft' competences, of the trainee commander, or indeed of the experienced commander. This area is the focus of the work of many of the psychologists who are investigating the nature of effective command.

That other and more complex skills come into play has never been in doubt. It has been recognised for a long time that successful commanders often appear lucky, Napoleon himself is said to have called for 'lucky generals'. He knew as well as anyone that his successful generals made their own luck, but he lacked the diagnostic tools that identified the essential ingredients to enable him to identify and select with confidence. As evidenced by many of the chapter authors in this volume, there is little changed over the two centuries dividing us.

Conclusion

To begin with, this chapter has attempted to set a foundation, mainly for the benefit of those unfamiliar with the terms, organisation and culture of the armed or emergency services, for the understanding of the terms and assumptions made in the following chapters. Thereafter, it can be seen that there are several issues in the field of incident command and critical incident decision making that continue to pose a challenge to those concerned with managing, training and selecting people for command roles. For example, how much can a commander's potential be assessed in advance? How closely linked are the skills of command and management? If some people are more effective in stressful situations than others, why is this and can the competence be acquired? What is experience and, if it is essential to the effective exercise of command, can surrogates for it be found?

At an organisational level other questions emerge. Is the distinction and relationship between the levels of command real or synthetic, or are they an inheritance from the times when communications were less sophisticated than now and a new model is needed? If the latter, should strategic level commanders feel free to intervene at tactical or operational levels now that technology increasingly permits them to?

As in many fields of endeavour, lack of a common language and conflicts of professional interest have the potential to cause some misunderstandings. It is entirely possible that some of the debate about command or management, or command and leadership, are academic questions in reality. However, there is no doubt whatsoever that the reason these questions are asked is valid. We need to find better ways of selecting and training commanders, and a better understanding of what command entails. If we do not achieve these understandings, we shall always be vulnerable to placing people in positions who, for their good and others', should not be there. That is the challenge.

Notes

1.   Burke & Hendry (1977) also found that in 81% of cases, no alternative courses of action other than the one chosen was considered.

2.   Colonel R. Sanders, former Chief of the Fire Department of Louisville, Kentucky, in a lecture to the 1st Brigade Command Course of 1994 at the fire Service College, Moreton-in-Marsh, UK.

3.   Plato, *Republic,* lines 415,a-d. (Any edition).

4.   Captain, W.W. Roskill D.S.C., M.A, F.R. Hist. S., Fellow of Churchill College Cambridge, Royal navy (Ret.) wrote 'The Art of Leadership', published by Collins, London, 1964: now out of print but available in some libraries, including that of the Fire Service College at Moreton-in-Marsh, UK.

5.   Brunacini A. (1985) p. 247. (This is just one example of a section of 'one liners' by the author exhibiting his widely respected wisdom in the field of incident command.)

6.   The author is grateful to B.L. Fuller, C.B.E., Q.F.S.M., for his observations, which have contributed to the author's understanding of these issues.

7.   Klein (1998) identified that fireground commanders make 80% of their decisions in under a minute, and also that they have '..little inclination towards systematic thinking. Instead they would make a gut choice.' (p. 4, 11).

8.   Hansard, 18th April 2000. Corporate Homicide Bill received a second reading in the House of Commons on 27th October 2000.

9.   *Fire Prevention*, 289, May 1996, pp. 4. (Judge Richmond Harvey QC held that Hampshire County Council, as the fire authority responsible for Hampshire Fire Brigade, was to blame for the destruction of a large computer firm's HQ after one of its officers made the decision to depart from normal procedure and turn the sprinkler system off before the fire had been brought under control).

References

Adair, J. (1968) *Training for Leadership.* London: Macdonald.

Brehmer, B. (1996), Dynamic and distributed decision making. *Journal of the Fire Service College, 1, 2*, 17–36.

Brunacini, A. (1985) *Fire Command.* Quincy, Mass.: National Fire Protection Association.

Buck, G. (1999). *The Role of Cognitive Complexity in the Management of Critical Situations,* PhD Thesis, University of Westminster.

Burke, E. & Hendry, C. (1997) Decision making on the London incident ground. *Journal of Managerial Psychology, 12*, 1, 40–47.

Gilchrist, I. (2000) *An Analysis of the Management of Information on the Fire Service Incident Ground.* PhD Thesis, University of Manchester.

Gordon, A. (1996) *The Rules of the Game.* Jutland and British Naval Command, London: Murray.

Graham, G. (2000) Ethical Decision Making. Paper presented at the 10th Annual Incident Management Symposium, Phoenix, AZ, 17th August.

Heller, R. (1994) Putting the Total into Total Quality Management. *Management Today,* 60

Home Office (1997) *Dealing with Disaster. (3rd Ed.),* London: H.M.S.O.

Home Office (1999) *Fire Service Manual, vol. 2, Fire Service Operations, Incident Command.* London: H.M.S.O.

Klein, G. (1998) *Sources of Power – How People Make Decisions.* Cambridge, Mass: MIT Press.

Lindley, S. (1999) *Decision Making Strategies among Traders in an International Bank.* MSc Thesis, University of Hertfordshire.

McCann, C. & Pigeau, R. (2000) (eds.) *The Human in Command: Exploring the Modern Military Experience.* New York: Kluwer/Plenum.

Von Clausewitz, C. (1976) *On War.* Edited and translated by M. Howard & P. Paret, Princeton: Princeton University Press.

3 Managing the Spike: The Command Perspective in Critical Incidents

Peter C. Sarna
Chief of Public Safety (ret.), East Bay District
Captain of Police (ret.), Oakland Police Dept.

Since December 1998 Chief Sarna has been a Principal in The Chief Executive Group, a public safety consulting firm in the San Francisco Bay Area. Prior to that, he served for ten years as the Director of Public Safety in a large, complex special district with responsibility for police, fire, and emergency medical services in diverse parklands and open spaces spread throughout a two-county jurisdiction in Northern California larger than the state of Rhode Island. Previously, he had served twenty years with the Oakland Police Department, from which he retired at the rank of Police Captain in 1988. He has experience commanding virtually every major division within a major metropolitan police agency and was the Oakland Police Department's lead trainer in the area of critical incident management. He also has had extensive hands-on experience as a front-line incident commander in a wide variety of major incidents, and has taught and consulted widely on leadership, performance management, strategic planning, and crisis decision-making with numerous local, state, and federal agencies. Chief Sarna has a B.A. and M.A. in Sociology from Boston University. In 1993 he was awarded a Fulbright Fellowship in Police Studies to examine approaches to critical incident management in British police forces.

www.chiefexecgroup.com
www.ebchief@aol.com

Persons above this lowest or operative level in the administrative hierarchy are not mere surplus baggage…even though, as far as physical cause and effect are concerned, it is the machine gunner and not the major who fights the battles, the major is likely to have a greater influence upon the outcome of a battle than any single machine gunner.
(Simon, 1976, p. 2)

As Chief Sharp [Assistant Fire Chief of the Coldenham, New York, Fire Department] entered the cafeteria to conduct his initial size-up, he found pieces of the roof dangling overhead. He noted the suspended drop ceiling was still intact. Making his way towards the rear of the room, he was met by the school principal who frantically pleaded for him to start CPR on a critically injured child. Sharp ordered a firefighter to start CPR as he continued to size up the disaster.
(Conboy, 1990, p. 86–88) 'Coldenham School Collapse Kills Nine Students'

Introduction

For the past thirty years, I have been fascinated, as both a practitioner and observer, with the process of incident command as I've watched it enacted both well and poorly in a wide range of events and settings. In more recent years, I've had the good fortune to share experiences and perspectives on incident command issues with counterparts from a number of different work domains, including air traffic control, emergency medicine, fire services in both the U.S. and Britain, commercial and naval aviation, and marine petroleum. These exchanges have often been facilitated by researchers in a number of academic centres (e.g. Bea, 1994: Flin, 1996; MacKenzie, 1997; Roberts 1990), who share a passion for bridging the worlds of researchers and practitioners, thereby informing both knowledge and practice. Each and every time representatives from a collection of work domains came together, it was not long before somebody, despite coming from markedly different cultural and technological worlds, shouted 'Eureka, it's all the same stuff'. The 'stuff' to which they were referring was, of course, incident command, which has only recently emerged in the states as a distinct discipline within policing. This chapter offers some insights into the development of incident command concepts and principles in U.S. policing, reviews some lessons learned from a number of major disasters (these are interspersed throughout the text, while more general lessons are summarized in the concluding section), and concludes with some thoughts on how the discipline must continue to evolve if understanding and mastery are to be achieved.

The Nature of Critical Incidents and the Organizational Response

Critical incidents – natural and man-caused – are life-safety events that generate extraordinary demands on police, fire, and emergency medical services. They exhibit tremendous diversity, ranging from large-scale chemical disasters to spontaneous sports celebrations. A quick review of typical U.S. incidents from February 1999 reinforces the extent of that diversity, in terms of incident scale, type, required response, and community of occurrence (shown in parentheses).

- Train crash injures 27 (Sacramento, CA)
- Police lob tear gas at rioting fans (Denver, CO)
- Natural gas leak forces evacuation of hundreds (downtown Seattle, WA)
- Explosives plant blast injures 2 (Hollister, CA)
- Officer shoots parolee after high-speed car chase (Manteca, CA)
- 28 arrested in raid at rave club (Tampa, FL)
- One snowboarder dead, one lost in avalanche (Mt. Baker, OR)
- Authorities try to coax Bay Bridge jumper down; traffic snarled (Oakland, CA)
- Tanker spills sulphur on Highway 4; residents evacuated (Concord, CA)
- Romanian riot police block march by protesting coal miners.
- Mercury scare at high school; more than 40 students treated for exposure (Phoenix, AZ)
- Rail car carrying hazardous materials explodes; nearby residents evacuated (Clymers, NJ)
- Odd standoff; man threatens suicide with a crossbow (Dallas, TX)
- Five killed in chemical plant blast (Allentown, PA).

In terms of definition, we define critical incidents as any condition, situation, set of circumstances, event, or occurrence that:

- Poses a significant life-safety threat, disrupts essential community routines or vital services, or requires unusual amounts of public safety resources to manage.
- Diverts resources from routine operations and requires activation of specialized resources.
- Demands a high degree of command-level coordination and direction (sometimes referred to as a command post or incident command system mode of functioning).

The jumping-off point for police disaster operations is always routine operations, the nature and organization of which differ sharply from critical incidents. This is implied in the last element in the above definition –i.e., something extraordinary or special is required to manage them. Hence, it is imperative that response organizations shift rapidly from routine to crisis operations to handle both the operational and management demands of an unfolding incident. During the initial response, management needs often are given short shrift because of the urgency of emergency tasks and the absence of supervisory and management personnel. Yet, time and time again, we have found that the speed at which this transition occurs is a major determinant of whether an incident will be managed effectively. The important role that first-line supervisors and commanders play in making initial situation assessments and initiating this shift in operational modes cannot be overemphasized. Some attention must also be paid to training first arriving officers to initiate the rudiments of incident command (e.g., setting perimeters, establishing a command post) and avoid actions, such as blocking traffic lanes with police vehicles, which may interfere with later operations.[1]

Day-to-day policing operations are organized to handle certain types of incidents efficiently and provide reasonable margins of safety for police officers and citizens alike. There are three basic challenges supervisors face in managing the dynamic relationship between resource levels and workload demands.

- The first involves allocating limited resources during periods of excess demand.
- The second involves managing the use of uncommitted resources during periods of slack.

- The third involves making rapid transitions and reallocating resources during emergencies (i.e., sudden up-shifts in the demand for resources caused by a single incident or multiple incidents).

The last scenario is the one in which we're interested for purposes of this chapter. It results when the routine of policing is disrupted suddenly by an incident that requires a major resource shift, which sharply reduces the ability of the department to continue servicing routine calls. 'Managing the spike' (something I'll talk about later) created by such incidents is a critical aspect of incident management at both the first-line supervisor's and incident commander's levels.

Making the above transition entails more than simply shifting resources to new work. The nature of that new work – dealing with crisis – is dramatically dissimilar from routine police work and, more importantly, its organization is radically different from the manner in which routine work is organized. In order to fully understand these differences, and therefore appreciate the challenge involved in making the transition between the two modes, it is necessary to examine the routine of policing closely.

During routine operations, whether the demand is generated by citizen calls – for service or by officers' self-initiated activities, policing incidents usually require minimal resource commitments (one or several cars), are generally completed in a brief time (30 minutes is a point of reference for many departments), and are not usually of a life-and-death nature. A normal work shift typically will consist of handling a succession of such calls or activities with varying amounts of 'free' time in between calls (the amount obviously varies by policing context). To service this routine, police work is decentralized, direct supervision of necessity is usually brief and infrequent, officers are taught to be self-reliant, semi-independent actors, who exercise substantial discretion in what they do, and work is loosely coupled – that is, the ongoing coordination required among officers is not particularly high. This working profile is almost the direct opposite of what is required in a crisis, where work must be centrally directed and coordinated, actors must function interdependently, and work is tightly coupled.

Critical incidents impact operations in a number of significant ways:

- Routine operations are disrupted as resources shift to handle the emergency. The shift in resources, largely personnel, can result from formal assignment or self-assignment of resources to the incident. The latter type can pose special problems in tracking and coordinating on-scene resources. The diversion of resources may leave routine events uncovered for long periods of time.

- Specialized resources, either from within or outside of the agency, are needed or are activated. These then have to be merged into operations upon their arrival.
- Organization and coordination requirements increase dramatically as the incident grows in size, scope, and complexity. These include:

  - Diverse types of resources may be required.
  - Large amounts of resources may become involved.
  - The duration of the incident may extend beyond a normal work shift, creating relief requirements.
  - Activity may have to be broken down into geographical and/or functional sub-areas (i.e., sectored or branched in the Incident Command System).
  - The points (interfaces) at which communication needs to occur may multiply, causing congestion and requiring a comprehensive communications plan, including a schedule of regular face-to-face briefings.
  - Multiple work spaces may have to be created and coordinated to manage different aspects of the incident-assembly areas, response points, exclusion zones, perimeters, witness debriefing areas, press staging areas, etc.

Because of extreme time pressures and high stakes, often involving matters of life and death, conditions of high stress will also prevail. At some point, high stress levels and the inability to cope with them will result in degradation of both information-processing abilities and decision quality. The situation is compounded by the fact that major incidents occur relatively infrequently. As one British police commander has observed, 'They are not the kind of incidents that occur on a regular enough basis to allow many incident commanders to build a personal data base of experience.'[2] Consequently, developing and maintaining both individual and organizational competency in incident management is a major challenge.

Thus, there are few natural opportunities to develop and regularly test critical incident management skills. Consequently, for inexperienced incident commanders, their first test will be similar to learning to swim in the deep end of the pool without assistance. This undoubtedly will hinder the rapid transition from routine to crisis operations.[3]

When examined closely, however, we see that there are broad management tasks – termed functional requirements – that need to be carried out to manage any incident, regardless of differences in their underlying nature (e.g., the differences between a hazardous material incident and a reported child abduction). Hence, managing one type of incident turns out to be good training for managing a variety of incidents, though the correspondence obviously will not be one-to-one. On the scale of 'micro' (e.g., responding to an in-progress residential burglary) to 'macro' (e.g., the 1989 Loma Prieta earthquake) incidents, functional requirements actually differ little. This enables us to learn the skills of incident management on, in effect, a range of 'mini-critical incidents'. These skills, especially if they are practised and refined constantly, are then transferable to far larger incidents.

The Command Perspective

The opening quotations underscore the importance of effective incident command in two very different settings – war and firefighting. In this regard, policing is no different. The first quotation reflects the fact that actors at different levels in an organizational structure add value to critical performances in different ways. They, in effect, bring different perspectives and skills sets to incident scenes. At the lowest level, task specialists perform a range of critical technical and tactical skills, while at higher levels supervisors and commanders add coordination, planning, and evaluation skills to the mix. Simon (1976) referred to the former as horizontal specialization and the latter as vertical specialization, which he saw as '...absolutely essential to achieve coordination among the operative employees.' (p. 9). Simon, however, was speaking in terms of a single organization. The coordination requirements in critical incidents, of course, are compounded greatly by the fact that they are often multi-organizational and multi-disciplinary in character.[4] A further complication is that first-arriving responders, by the nature of their jobs and their training,[5] focus automatically on task requirements, e.g., suppressing the fire, rescuing the injured, dispersing rioters, rather than on coordination needs. These markedly different orientations are depicted in Figure 3.1.

Figure 3.1 Incident requirements

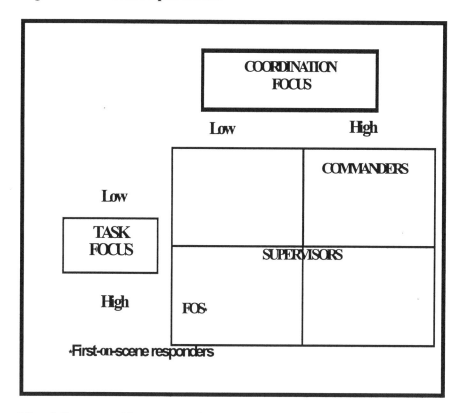

The failure to address critical incident management needs in the early minutes of a rapidly unfolding incident may create problems, such as blocked ingress and egress lanes, no control over volunteers, duplication of effort, and poor utilization of on-scene resources, later on. During the early minutes of the July 1981 Hyatt Regency Hotel structural collapse in Kansas City, Missouri, which killed and trapped dozens of party goers, police supervisors and commanders rushed to the scene, only to become quickly involved in ground-level work, i.e., rescue and medical aid. The result was that critical incident management functions, such as keeping ingress and egress routes open, were delayed. This complicated and interfered with later operational needs. As the department concluded in its after-action report, '...police commanders must look beyond the initial response of police personnel and take the necessary actions to provide for the long-term sustained management of an emergency situation'.[6] The task of shifting quickly into a crisis mode of operations is further complicated by the structure and routine of policing during non-crisis periods, which I discussed earlier.

The description of Chief Sharp enacting the role of incident commander is both poignant and prototypical. It epitomizes the command perspective, central to which is the ability to attend to critical aspects of an unfolding situation – which implies that other aspects are ignored or quickly dismissed as less important – make sense of what's going on, determine what needs to be done about it, and then execute a course of action to stabilize the situation and eventually set things in good order. Getting in role, and staying in role, thus, are essential, as failing to do so diverts attention from critical requirements and consumes valuable time better spent on sense making and course-of-action (COA) generation. As one observer put it to me, 'There's a 1,000 things happening, you're aware of 100, and you can only do something about 10!' Thus, the ability to perceive, understand, and focus on a few key aspects of an unfolding incident is key to performance as an incident commander. As Simon (1976) notes, this type of work is especially important if wars, or their equivalent in other work domains, are to be won.

There appear to be two opposite notions of incident command competencies: the first sees incident command as merely an extension of everyday management competencies; the second, in contrast, views incident command as a distinct discipline with its own set of specific competencies. The second strikes me as the more useful, though there probably is substantial overlap between the two sets of competencies. In fact it may be the conditions of performance that truly distinguish the two. The following list describes these competencies, along with significant conditions of performance, which are essential for effective incident command performance:

- Handling multiple, demanding problems concurrently under conditions of high stress and emotion, urgency, confusion, and uncertainty.
- Ranking different, competing tasks in order of importance in compressed time frames with limited resources.
- Knowing and executing predefined options for handling certain types of crises and their impacts.
- Formulating new courses of action to resolve or mitigate novel crises and their impacts.

- Coordinating the activities of numerous specialists, each having to some extent, conflicting decision premises, mental models, and views about needed courses of actions, and a vested interest in seeing their model implemented.
- Conducting accurate situation assessments and ordering actions that will result in major social and economic dislocations.
- Making high-risk decisions with life-and-death outcomes.
- Activating an appropriate level of organizational response to handle the crisis, yet maintaining normal operations at the highest level feasible.
- Coordinating actions and negotiating agreements across organizational and disciplinary boundaries.
- Managing personal stress levels to prevent or minimize performance degradation.

Organizational Capacity/Capability and Risk

Incident management capabilities among U.S. police agencies vary widely, and frequently depend to a large part on size (i.e., the sheer ability to field large amounts of manpower with little or no delay is certainly an important dimension of effective response), which is generally a good index to the frequency with which the agency will be required to handle a variety of incidents. Obviously, incident frequency is a factor in raising experience levels and building incident management competency within an agency. However, large size in and of itself is no guarantee that an agency will perform effectively in the face of crisis, as attested to by a number of major failures in recent years, including the Branch Davidian standoff in Waco, Texas, the post-verdict Los Angeles riots in Spring 1992, and, more recently, the World Trade Organization-related riots in Seattle, which culminated in the firing of the police chief for failure to plan adequately for violent demonstrations.

Figure 3.2 is an attempt to show the relationship between organizational capacity (which is a question of how much)/capability (which is a question of how well, or competency) and the risks associated with critical incidents. It has proven to be a useful analytical tool to identify and understand key factors involved in incident command failures.

The diagonal axis in figure 3.2 shows the relationship between risk level in an incident and organization/capability, which I've shortened to capacity for this discussion. The scale for all of the axes in the diagram –

vertical, horizontal, and diagonal – proceeds outward from the center, from low to high in all directions.

Figure 3.2 Organizational Dimensions

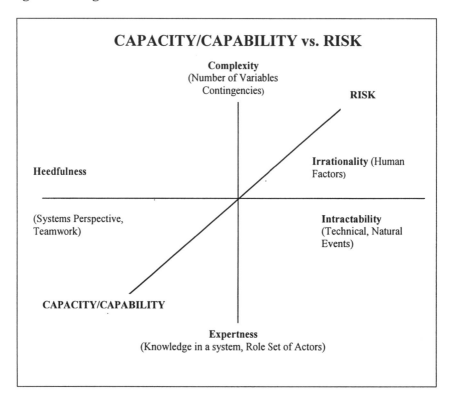

The risk level in any incident consists of two sub-dimensions: complexity, which refers to the number of variables and problems that require management, and intractability, which is defined as the extent to which a responder's actions are likely to influence or control causal factors (e.g., a wild-land fire, tornado, deranged gunman). The area created by joining the two plots together is a graphic representation of the extent of risk (which is also a measure of the likelihood of partial or complete failure) in a given incident. For example, most weather-related disasters would score high, as would the Branch Davidian standoff in Waco, Texas in 1996. The 1991

Oakland Tunnel Fire in California, which destroyed several thousand homes and killed 28 persons, rated high on both dimensions. From the time of ignition to the time the wind-driven flames began to die down, the fire service was chasing the incident demand curve. Tragically, we seem to relearn old lessons every several years: whenever the underlying causal agent in an incident is highly intractable (i.e., won't succumb to the will and efforts of human actors), more time and effort should be spent in prevention and mitigation efforts.

The opposite lower left-hand quadrant depicts organizational capacity, which is a function of two factors: expertness (know-how) and heedfulness (Weick & Roberts, 1993). The second factor is critically important in a system of multiple, spatially distributed actors whose specialized roles are characterized by high interdependency e.g., Special Weapons and Tactics (SWAT) team members and hostage negotiators. Heedfulness refers to the extent to which actors are mindful of critical interdependencies and therefore take steps to ensure shared understandings and coordinated actions. Plotting values for these two variables and connecting the two data points yields a measure of total capacity. The resultant area reflects the amount of capacity an organization brings to a particular scenario. Obviously, worst-case scenarios combine high-risk factors with low organizational capacity for managing and resolving incidents. Unfortunately, it is commonplace to find high expertise and low heedfulness operating simultaneously in many response organizations. The Federal failure at Waco was plagued by low heedfulness among key response elements. It has been routinely identified as a causal factor in a wide variety of accidents, typically under the rubric of 'human factors'. Because it has been repeatedly implicated in incident command failures in cockpits, on ship bridges, in police and fire command posts, in hospital emergency rooms, and in numerous other crisis management 'hot seats', to use Rhona Flin's (1996) metaphor, it must be one of our highest research and training priorities.

Managing the 'Spike'

For years I've referred to incident command as 'spike management' in order to capture the twofold impact that incidents usually have – one on individual functioning, the other on organizational functioning. All organizations have steady states, or normal operating routines, configurations and tempos (realistically these fall into some sort of range with upper and lower limits that define normalcy), which they try to preserve and protect through risk management strategies and programming. So, too, do individuals, though their protective apparatus is less elaborate. Disruptions to these steady states or routines can trigger classic stress responses, particularly in untrained or inexperienced responders. The response, which is often described as 'going into adrenaline overdrive', can produce both functional (e.g., mobilize the body to deal with a severe threat) and dysfunctional effects (e.g., result in loss of situational awareness).[7] To the extent that the effects are dysfunctional, both ground-level task performance and incident management, including critical cognitive work, such as perception, information management, and decision-making, will suffer.   Thus, training simulations that mimic incident stressors (e.g., overload, ambiguous information, and briefing higher-ranking staff) should be used regularly to prepare incident commanders, particularly novices, to manage performance-degrading stress effects in themselves and others effectively.

Incident commanders are also taught to remain alert for stress-related effects in on-scene personnel, especially with the passage of time and increased exposure to traumatic conditions.   This function may be delegated to another position within the incident command structure, such as the safety officer commonly found in fire service iterations. One of the key functions of command is to counteract or mitigate stress-related effects that might cause degradation of task performance within responder groups and the command element. The operative phrase, particularly in the early, hectic moments of an incident, is 'calm is contagious'. Hence, an incident commander's composure, in both word and deed, is a critical factor in maintaining calm among responders and other affected persons.[8] In some cases, incident commanders may have to remove workers temporarily (regular rotations between work and rehab modes is the preferred arrangement) from the incident-related duties for rest and recuperation, or permanently, because of fatigue or other considerations that impair job performance. In all cases, incident commanders must ensure that some meaningful form of post-incident stress debriefing takes place for both business and humanitarian reasons.[9] This has become standard practice in most U.S. police agencies.

The organization undergoes an analogous stress response, albeit on a much different level. As the organization mobilizes, there is a massive, often uncoordinated, convergence on the incident scene. The first wave of response usually consists of a surge of police, fire, and emergency medical vehicles, which must be coordinated quickly if key access points and routes are to remain open.[10] Following this first wave, the primary focus of the organization, along with all sorts of specialized resources from within and without the agency, shifts from the operating routine to the crisis. How long the organization can or should sustain both that focus and level of resource commitment will be an ongoing question. In many instances, this shift is governed by predefined response protocols, which organizations have formulated to manage events within their crisis portfolios. These protocols usually address how the organization will manage adverse impacts on normal business operations, as there are few contexts in which normal business operations can be suspended for significant periods without adverse impacts. For this reason, incident commanders and uninvolved staff must at some early point evaluate whether the organization has the ability to service its routine needs fully, partially or minimally. A dual management structure is also usually required, with one manager focusing on handling the crisis, while another manager oversees the continuity of routine operations. Generally, if a single commander is forced to fill both roles for an extended period, one or both operational areas will suffer. Whatever the setting, be it a prison, school, production plant, hospital, or shopping mall, continuity of normal operations will become an issue in the course of managing a major incident.

As significant resources are shifted from routine operations, in policing this may necessitate reconfiguring field operations. This typically means reducing field staff to skeletal levels and raising the service threshold to respond only to emergency calls. Other routine activity must be placed on hold until the incident is resolved, or until sufficient resources are freed up from the incident or made available from other sources to resume normal business activity. The resultant work backlog may create a secondary crisis at some point.[11] With respect to the convergence phenomenon, the incident commander has two major concerns: first, he must ensure that adequate resources are allocated to the emergency; and, second, he must also ensure that some provision is made for managing the disruption in routine operations. In larger agencies another commander usually assumes responsibility for the continuity of routine operations, though in smaller ones incident commanders may have to shoulder both roles for a time. Once the incident is under control, incident commanders need to evaluate and adjust resource commitments. In some cases, significant resources can be freed up from the incident and re-integrated into normal operations, though this should be an orderly, supervised, and systematic process.

The Nature of Critical Incidents: A General Model

Up to the early 1980s, it was typical of police departments throughout the U.S. to maintain inventories of hazard-specific plans.[12] For instance, the Oakland Police Department (located across the bay from San Francisco) in 1982, by actual count, had in excess of twenty specific crisis plans. These ranged from plans for managing earthquakes and hillside fires to a plan for handling extremely rare snow emergencies in the Oakland hills. Each demanded constant revision and updating, which few agencies were able to do. Hence, they were often not regarded as reliable guides to decision-making, forcing inexperienced supervisors and commanders to muddle through whenever actual incidents did occur.

Field commanders also found that, voluminous plans in massive three-inch, three-ring binders were difficult to access in fast-breaking emergencies, especially under often wretched weather and scene conditions (a common observation was that they had more utility as doorstops!). As we learned quickly, a pocket cue card with five to seven basic actions to take in the first hour of any event proved a far more practical and useful performance aid for folks at the 'sharp end of the stick'.[13]

As a result of the deficiencies of the 'a-plan-for-every-disaster' approach, single, all-hazard plans, incorporating Incident Command System concepts and tools, have become the norm. These plans are based upon experience and research that show that seemingly very different incidents share common features and similar management requirements. There are three major premises underlying the general model of incident management that is now taught to incident commanders in most California and U.S. police departments. The first is that there is a common set of functional requirements that exist across a range of incidents (these coincide very generally with the major functional entities within the Incident Command System). The second is that there are discrete stages within incidents, during which certain functional requirements tend to predominate. The third premise is that certain management activities, particularly situation assessment and course-of-action generation and evaluation, have to be repeated continuously ('repped' in the vernacular) to ensure accurate sense making and appropriate course-of-action selection.

Incidents Have a Common Anatomy

> Commanders need to realize that all tactical situations have the same basic elements and, therefore, they can apply a standard approach to them.
> (Brunacini, 1985 p. 5).

We have found a simple, easy way to conceptualize these premises in the classes we teach – we ask students to think of incidents as 'donuts'[14] (yes, the kind that you eat; we thought that this analogy would resonate particularly well with cops, given the almost genetic affinity between them, coffee, and donuts!) The 'hole' of the Critical Incident Management (CIM) donut, shown below in Figure 3.3, is referred to by a number of terms – the hot zone, zone of action, or emergency scene. In it exists some type of technical problem, usually posing some sort of grave risk, requiring specialist responders and solutions. Around the hot zone, in the 'meat' of the donut, as it were, lies another sphere of activity – scene management duties, which is the term of reference in pertinent California legal codes defining what service has lead authority for different types of incidents, or more generically, incident command functions. At a high level, these functions are virtually identical in a wide range of incidents, including major crimes, riots, chemical spills, road races (not all critical incidents are negative), major fires, earthquakes, floods, search and rescue operations, and community festivals. Obviously, at ground level these are very different events, though they are managed in largely the same manner.

Perhaps the simplest way to demonstrate the existence of broad functional requirements across a wide range of incidents is to examine a sample of incidents, as we do in the basic incident command course that we teach.

Figure 3.3 CIM 'Donut'

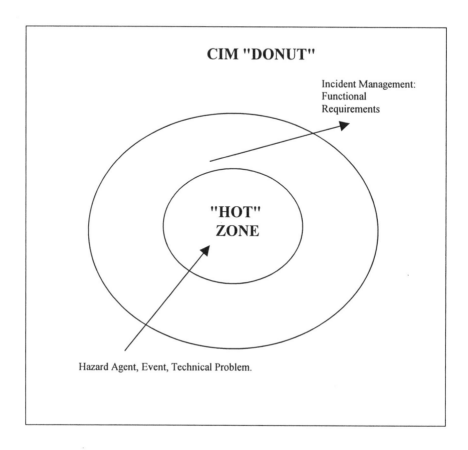

For instance, we share a number of cases studies with students, identify the functional requirements of each incident as a group exercise, and then determine requirements that are common to all of them. We array the lists of functional requirements for the cases in the sample side by side on a work sheet. The degree of overlap jumps out quickly. In short order, students realize that there are a set of basic management functions that reappear in almost every case studied. These are then reduced to a pocket memory aid or checklist. (Another useful exercise is clipping a week's worth of news articles on critical incidents and analyzing them for commonalities. This is an extremely simple, inexpensive, and pain-free form of learning from other agencies' disasters.).

As an example, one of the highest priority functions is what we term access control, which encompasses a number of sub-functions (bundling related items under a major heading is an effective strategy for recalling important information under stress; in this sense, it is similar to the use of mnemonic devices). Regardless of the nature of the incident, incident commanders are almost always confronted with the immediate need to establish control over an affected piece of ground, be it a neighborhood, event venue, or single room in an apartment complex. This is usually accomplished by setting cordons or perimeters, controlling entry and exit through established checkpoints, defining specific use areas within or adjacent to perimeters, and so on. All of these actions are intended to contain threats (e.g., a deranged gunman, a chemical leak, or crowd of rioters) within a defined area where controls can be applied, protect innocent third parties from harm, eliminate or minimize interference with emergency responders, and facilitate movement into and out of the affected area. Other major functional requirements include media relations, command post operations (particularly the 'who's in charge' issue in multi-agency operations), situation assessment, logistics, communications, liaison, and continuity of operations. Depending how you bundle sub-functions under major headings, there are eight to ten functional requirements that incident commanders have to remember, though not all may come into play during any single incident. These are easily codified in a pocket cue card, or checklist if greater detail is desired. The Incident Command System interestingly also functions as a checklist of important incident management functions, though it was intended primarily as an organizing tool.

Incidents Unfold in Discrete Stages

The sequence in which incidents unfold and response takes place is not random. In fact, it can be broken down into discrete stages that provide a framework for the conduct of incident command work. The model in Figure 3.4 consists of five stages, including pre-incident and post-incident stages.

Figure 3. 4 Incident Stage Model

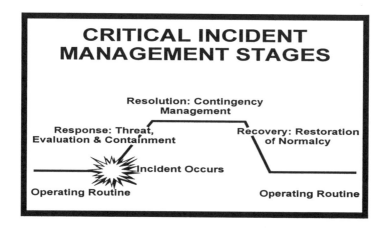

Different types of work are called for in different stages. Preparation should obviously take place in the pre-incident stage, while formal critique is usually an important activity in the post-incident stage. The objectives of each stage are shown in Table 3.1.

Table 3.1 Incident Stages and Objectives

STAGE	OBJECTIVES
Normal Operations – Pre-incident	Prevention, Mitigation, Preparedness
Response and Containment	Assessment, Stabilization, Initial Counter-measures, Shift to ICS Mode
Contingency Management	Management, Resolution of Problems
Recovery	Restoration of Normality
Learning* Post-incident	Improvement of Capabilities

*Labeled as such only for analytical purposes. Obviously, learning should and does take place during all stages. Learning is what makes critical mid-course adjustments possible. However, at this stage; learning is formally addressed to ensure that lessons are codified and disseminated in training, policy statements, and formal case studies.

Models enable us to rationalize the incident command process, so that critical tasks can be attended to and carried out in a logical sequence. The sequence, though, should be seen as a guidepost and not enacted as invariant. For example, subsequent to the collapse of a double-decker freeway in Oakland, California during the 1989 Loma Prieta Earthquake, the police department quickly sealed off the surrounding area. Inadvertently, they also blocked access to hundreds of small and medium size businesses in West Oakland. Normally, re-establishing access would not have been a concern until rescue and body recovery work was fully completed at the site. After all, that is when it's usually done in the standard sequence of activities. However two things became clear immediately: first, rescue and body recovery operations would likely span several weeks; and, second, businesses marooned inadvertently by security barriers could not survive business disruptions beyond a few days. Another crisis of major dimensions was smoldering as smoke still rose from the collapsed freeway. Fortunately, as a result of strong connections with the business community and the foresight of several police commanders, a program was set up quickly to restore limited business-related access to the affected area. This facilitated business recovery literally in the midst of a major disaster, averting significant economic losses to the city, the community, and area businesses.[15] Had the police commanders in question followed a perfectly linear model, re-establishing access to the affected area would have been delayed until rescue and body recovery operations were completed. Doing so would have unquestionably precipitated a secondary crisis of major dimensions. Fortunately, they remained flexible, maintained situation awareness, and engaged in some 'down board' thinking.

Continuous Situation Assessment is Essential

Failures to revise often produce ugly surprises. (Weick, 1995, p. 13).

Fast-moving episodes are punctuated by intervals which provide opportunity for reflection. (Schon, 1983, p. 278).

I have written elsewhere on the role of sense making in incident command – what I term developing and maintaining 'the big picture'. In this regard, though I believe we've got the first two elements of the incident command trinity – function and structure – fairly well in hand, I remain deeply concerned that we don't truly understand the nature of the cognitive work that is central to incident command performance. This is the process piece.

It is what Crichton and Flin, (this volume), Klein (1998), Roberts et al. (1994), and a host of others in a variety of work domains, are trying to figure out.

The gap that I perceive in knowledge and practice is particularly troublesome because flawed process, in my judgment, in great part accounts for the collapse of sense making in a number of recent tragic failures – the loss of 14 firefighters in South Canyon, the deaths of 4 firefighters in a Seattle warehouse fire, the fire and resultant loss of life, including numerous children, in the Branch Davidian standoff, and countless other incidents. I fear, in Schon's (1983) words, that we are regularly missing critical opportunities for reflection in the course of managing serious threats to life and limb. 'Going mindless' is not an option, in my judgment (see Langer, 1989).

An example from my personal experience underscores the concern I have about practitioners mistaking function and structure for process.

> As the simulation unfolded, the IC (Incident Command) assembled a briefing of key persons in the response organization and proceeded to ask a series of questions about the situation and the state of the response organization. An exercise controller quickly intervened and advised that the simulation was getting off track, i.e., the IC needed to focus his attention on staffing all of the positions in the response organization (ICS) before conducting the briefing. In effect, structure was assigned a higher priority than sense making. The tension lasted throughout the simulation. (Personal quote)

I obviously was the IC in the brief anecdote. Whereas I, and others, including the city's veteran fire chief, who was part of the core incident management team, got ahead of the exercise curve because the cues presented were part of a familiar pattern, exercise controllers tried to rein in the group and get it to focus on building the response organization. I saw this as a diversion of both attention and effort that in a real-world scenario would exact a grave penalty. I later explained to the exercise design group that they should have either given the group leeway to think and act 'down board' (appropriate outcomes could have been built in to penalize wasteful digressions if their Situation Assessment (SA) was inaccurate), or assigned role players with far less experience. Unfortunately, had they done the latter, I fear that novices would have learned the wrong lesson. At no point before, during, or after the exercise, was there any discussion of cognitive work, particularly SA.

Karl Weick, who has written an incredibly insightful analysis of the South Canyon wild fire that killed 14 smokejumpers several years ago (Weick, 1995), has suggested that cognitive processes should be regarded as the static elements within a decision-making, problem-solving system. In his schema, structural elements, such as positions and functions within ICS, would be regarded as dynamic elements, changing as situational needs dictated. This in fact was one of the original assumptions underpinning ICS, i.e., it was a modular, readily adjustable organizing framework. Weick's point represents a fundamental shift in mindset, particularly for those practitioners who have over-learned ICS as a management tool. We need to examine and consider his point closely.

If cognitive work is indeed critical to effective performance as an incident commander, there is a pressing need to identify discrete behaviors, skills, and tools that are essential for success, particularly under conditions of urgency, high stress, grave threats, and uncertainty. Over-reliance on structure, which often masquerades as knowledge and process, is certainly a trap we have to avoid. For you see, the question is not whether deck chairs should be rearranged on sinking ships; it is more a matter of when and under what circumstances. That is a far better question. And, the answer is 'it depends', and knowing 'upon what' is the essence of incident command.

Final Points

- Response organizations, particularly those dealing with recurrent events, are highly procedural. Thus, they are inherently flawed to deal with dynamic, novel, and equivocal situations. It may be difficult to counter this fundamental tendency to enact set routines after cursory sense making efforts, and shift such organizations into an action – learning mode, because most of their work is in fact highly routine and procedural in nature. They may be afflicted, thus, with a 'trained incapacity' to act in other than a tactical, procedural manner. Incident commanders must be taught cognitive strategies and skills to overcome this tendency.
- Specialized responders arrive with predetermined mindsets that often act as perceptual blinders to new or discrepant information, and thereby limit 'big picture' thinking. It is the role of incident command to ensure that a shared understanding of the situation is explicit and guides

decision making throughout a distributed system of actors. The lack of shared understanding increases the likelihood of uncoordinated or conflicted action. Incident commanders must have the skill sets for understanding and accomplishing cognitive work in a group of diverse actors who are often strangers. Otherwise, sense making – the foundation for all action – will fail or fall short of the needed level of understanding.

- Irrespective of structural features, cultural norms may impede important cognitive behaviors that are crucial to sense making and problem-solving (e.g., advocacy and inquiry, migrating decision authority). This is one of the central tenets of Crew Resource Management/Leadership, as taught and practised in commercial aviation (see Lodge; Crichton & Flin, this volume). Although assumptions, usually tacit and therefore untested, underpin all sense making and are thus important drivers of behavior, cultural norms, particularly in hierarchical structures, may act as barriers to making them explicit and subject to open testing. Again, incident commanders must be taught cognitive strategies and skills to overcome such barriers.

Notes

1.   As noted in the text, an axiom of incident management is that the speed at which a response organization shifts from routine to crisis operations will have a direct bearing on how well it manages the incident. Impeding this shift, however, is the fact that initial responders typically are more task – than coordination-orientated, and usually lack training in or sensitivity to higher-level incident management requirements. This inevitably delays the required shift in operational modes. Training initial responders in the rudiments of incident command can in great degree remedy this gap.

2.   R. Ramm, 'The Importance of Siege Mentality,' *Police Review*, May 11, 1990, p. 11.

3.   In April 1989 a relatively inexperienced South Yorkshire Chief Superintendent, David Duckenfield, according to the official report, failed to prepare for or react properly to a worsening crowd situation that eventually led to the deaths of 95 football fans in Hillsborough. Duckenfield, who appears to have been 'thrown into the deep end of the pool', after 23 years in criminal investigation, eventually retired for medical reasons stemming from the tragedy. See 'The Hillsborough Judgment,' *Police Review*, August 11, 1989, pp. 1612–1613.

4.   Bureau of Justice Statistics survey in 1997 identified 13, 540 local police agencies, 3,088 sheriff's departments, 49 state police agencies, and over 2,000 special police

departments in the U.S. The survey did not include federal police agencies. Many of these agencies have overlapping or neighboring jurisdictions with respect to responding to major emergencies. Hence, most large-scale U.S. incidents quickly turn into multi-jurisdictional incidents. This increases the complexity of the response and coordination requirements, and often results in fragmented efforts.

5. In July 1995, while teaching a seminar on critical incident management at the University of Portsmouth, a young Metropolitan Police constable shared a story with the class that illustrates the importance of teaching incident command basics to initial responders. On a late evening tour, as he explained, he and several other constables responded to a bombing incident in a downtown London shopping district. Upon his arrival, he started ticking off in his mind the basic tasks that an incident commander would have to address when he got on scene. He quickly realized that many of the tasks, such as scene protection, would not tolerate delay of any degree. This led to the further realization that, until a senior officer arrived, he was in fact the on-scene incident commander; without further delay, he commenced to take action on the mental checklist that he had created.

6. R. Mathis, R. McKiddy, and B. Way, 'The Management of Crisis', *The Police Chief*, November, 1982, pp. 48 –52.

7. As Patrick Lagadec notes, '…the degree of preparation has a direct effect on stress levels: when there is no prior experience, no tools, and no training, the organization [and the individual have] no repertoires of answers available, and this heightens the stress. The more unfamiliar the event, the greater the adaptation required. ' In P. Lagadec, *Preventing Chaos in a Crisis* (Maidenhead: McGraw-Hill, 1991), p. 65.

8. The calm exhibited by United Airlines Captain Al Haynes, during a major in-flight emergency in July 1989, is cited frequently as an example of command behavior that induces calm and discipline in others (see A. Haynes, *Flight Deck*, 3, 5– 21).

9. See R. Conroy, 'Critical Incident Stress Debriefing,' *FBI Law Enforcement Bulletin,* February 1990, pp. 20– 22, and C. Casey, 'Disaster Debriefing,' *Police Review,* June 2, 1993, pp. 14– 15.

10. The convergence of field units to an incident is nowhere greater than in response to the sudden 'announcement' of a high-speed vehicle chase. Hence, it is critical that field supervisors establish 'presence' immediately (typically over the police radio) to counteract the adrenaline rush, particularly among younger officers, that impels pursuers to take unreasonable risks and make poor decisions. This is what I refer to as 'managing the spike.'

11. Mutual aid arrangements exist throughout the U.S. to ensure that public safety resources are shifted quickly from areas of low need to areas of high need in a major emergency. These arrangements mainly encompass police, fire, emergency medical, and coroner services. However, they are expanding rapidly to include a variety of other disciplines, such as planning and engineering specialists, public works, and public utilities.

12.   I've argued for years that the best 'crash' course in critical incident management is the morning newspaper and the 6:00 p.m. television news. Within a week, the observer quickly appreciates the diversity of incidents occurring regularly in communities across the country and around the world, and understands the basics of incident management, even if bad examples tend to predominate.

13.   The British Transport Police have ingrained the mnemonic CHALET into the minds of line officers. It stands for the key steps that initial responders are to take: determine casualties, identify hazards, determine access routes, provide a precise location, specify the emergency services required, and determine the type of incident.

14.   Another useful analogy is an event venue, such as Wembley Stadium or Madison Square Garden. What occurs on the event floor, the 'hot zone', as it were, varies from night to night – a sports show one night, a musical concert the next, and a tractor pull the third. These are very different events, much like a hostage incident, lost child, or chemical spill are markedly different events. However, the management activity that takes place around the floor events – or the critical incidents described above – is similar, almost identical, across events, for example, getting people into and out of the venue, dealing with unruly or lost patrons, and managing lost or misplaced property.

15.   A. Hare, R. Nichelini, and P. Sarna, 'Recovery in the Midst of Disaster-Managing Access to the Cypress Freeway Collapse, ' *The Loma Prieta, California, Earthquake of October 17, 1989: Societal Response,* undated paper.

References

Bea, R. (1994) Management of human error in the operation of marine systems. *Marine Technology Society Journal,* Spring.

Conboy, M. (1990) Coldenham school collapse kills nine students. *Firehouse,* July, 86– 88.

Flin, R. (1996) *Sitting in the Hot Seat: Leaders and Teams for Critical Incident Management.* Chichester: Wiley.

Klein, G. (1998) *Sources of Power. How People Make Decisions.* Cambridge, Mass: MIT Press.

Langer, E. (1989) *Mindfulness.* New York: Addison Wesley.

MacKenzie, R. (1997) Introduction. Paper presented to the Critical Incident Management Conference. University of Portsmouth, September.

Roberts, K. (1990) Some characteristics of high reliability organizations. *Organization Science, 1,* 160–177.

Roberts, K., Stout, S., & Halpern, J. (1994) Decision dynamics in two high reliability military organisations. *Management Science,* 40, 614–624.

Schon, D. (1983) *The Reflective Practitioner. How Professionals Think In Action.* New York: Basic Books.

Simon, H. (1976) *Administrative Behavior. A Study of Decision Processes in Administrative Organizations.* (3rd ed.) New York: Free Press.

Weick, K. (1995) South Canyon revisited: Lessons from high reliability organizations. *Wildfire, 4, 4,* 54–68.

Weick, K. & Roberts, K. (1993) Collective mind in organizations: Heedful interrelating on flight decks. *Administrative Science Quarterly, 38,* 357–381.

4 Incident Command Functions

Alan Brunacini
Chief of the Phoenix Fire Department, Arizona, USA

Alan Brunacini joined the Fire Department in Phoenix, Arizona, a city with a population over one million, in 1958, and has been Fire Chief since 1978, leading a department of over 1400 firefighters. He is a 1960 graduate of the Fire Protection Technology programme and earned a degree in political science at Arizona State University in 1970, followed by a Master of Public Administration degree from Arizona State in 1975. Chief Brunacini is a popular speaker and author on fire service topics. His book, Fire Command, has become an essential text for students of fire fighting everywhere, and formed the basis of command systems now in use throughout the USA and other parts of the world. He has recently completed a second book entitled Essentials of Fire Department Customer Service. Alan Brunacini has served the National Fire Protection Association as chairman of various technical committees and also as Chairman of the Board of Directors. He has been recognised for his accomplishments with numerous awards and honours, including in 2001 the prestigious Paul C Lamb award from the NFPA.

My initiation to the incident command process started soon after I became a member of the Phoenix Fire Department in 1958. I was assigned to an engine company in downtown Phoenix for my first 12 years on the job. During that period I served as a firefighter, driver, and company officer. My initial occupational challenge involved developing an understanding of what was required to be a fire company team member. As a young firefighter I concentrated on learning the technical skills involved in doing the task level manual labour of firefighting. I recognised early in my career that effective firefighting operations contain an interesting contrast. While the basic work objectives are pretty simple (rescue/extinguish/conserve), the manipulative skills required for effective fire control are complicated and must be co-ordinated and performed quickly, many times by multiple work teams, in a sequential manner with lots of simultaneous activity going on all at once in different places. The point of this integrated firefighting effort is to create a directed, concentrated, operational response. The outcome of the response is also pretty simple − if the firefighting effort is bigger than the fire, the firefighters win. If the fire is bigger than the operational effort, the fire wins.

Fairly early in my career, I noticed that our company officer could effectively command and control our activities by direct supervision. We all lived together (literally) when we were on duty. Our company officer boss had continuous access to the crew. He always knew where we were and what we were doing. If he lost track of us, he came looking for us (on and off the fireground). He could quickly evaluate the effect of our action because he was right where that action occurred. He was an active participant in everything we did. If our company had the capability to solve a problem by itself, the communications between our officer and the crew was short and sweet and we were able to use our resources to get the job done. Activities that involved just our company typically went well, however, when we became involved in larger incidents that required the response of multiple companies a somewhat different process occurred.

In those days the Phoenix Fire Department was divided up into four battalions. Each battalion was commanded by an on-duty Battalion Chief (BC). The Battalion Chiefs were the bosses of the companies. They did administrative work, co-ordinated department programs and responded to fires to provide on-scene command. My station was located in the first battalion, but frequently responded to multiple unit events that occurred in the second battalion, so we routinely worked under the command of both Battalion Chief 1 and Battalion Chief 2. The two officers could not have been more opposite in their personalities and their approach to managing incident operations. Battalion Chief 1 was a traditional, 1960s type fire officer. He was an assertive, autocratic, intelligent and very experienced (30+ years) micro manager. Listening only occurred on those rare occasions when he wasn't talking (actually lecturing). Battalion Chief 2 was a younger guy, sort of quiet, laid back with a good sense of humour. He treated people with a 'light hand' and did not over manage. When our company responded to the 1st battalion, we turned right. Responding to the 2nd battalion required a left turn. Making those turns responding to a multiple unit structural fire incident was like travelling to a different planet.

A fire situation not controlled by about the first two attack lines quickly begins to reach a scale and complexity that requires an integrated collective effort among fire companies. These efforts must all occur within quickly developed and executed incident action plan (IAP). These larger situations require that multiple points of operation be quickly set up and that individual company efforts become integrated and co-ordinated to produce an adequate, well placed concentration of operational action. These fires require a level of co-ordination and integration that quickly exceeds the capability of individual fire companies operating independently. The size and complexity of these events requires an incident commander to be in attendance and be highly conscious. That incident commander must create and manage an overall game plan that connects operational action to incident operations. How this command process actually occurred during my fire company days depended (simply) on if we turned right or left. When we turned right, we all knew we were going to spend some exciting time with Battalion Chief 1. Upon his arrival he would abandon his vehicle, circle the incident at a dead run, screaming conflicting orders to anyone and everyone he encountered. His typical response to an expanding fire was to order the first wave of responders to inside positions with hand lines, call for more resources and then assign the second wavers to outside positions where they would blast the insiders with deck guns. The result of this lack of effective command in Battalion 1 was that we burned up a lot of property that was saveable and unnecessarily beat up a lot of firefighters during those operations where Battalion Chief 1 was circling and screaming like a hyperventilating lunatic.

Hanging a sharp left and taking a trip to the 2nd battalion produced exactly the opposite experience. Battalion Chief 2 would arrive, get out of his car and lean (his body language always looked sort of casual) against the door. He would direct companies to come to his vehicle (no portable radios in those days) where he would assign them face to face. Many times he would calmly engage the companies by asking 'whatdaya think?' – then he would actively listen, and together he and the assignee would develop and agree on a work place/function. He would never raise his voice and generally would smoke his pipe (another casual signal) as he evaluated conditions and observed the effect of operational action. He always had a clipboard with an old-time yellow pad and as he made assignments, he would note with a stubby little #2 pencil the who/where/what of their task. I never saw him leave his vehicle while a fire fight was under way and I never saw him get excited.

Predictably, both officers had a profound effect on my career. Both served as mentors, one negative and one positive. I was promoted to Battalion Chief at 13 years and served in that position for 5 years. As a Battalion Chief I always attempted to imitate BC2 (minus the pipe, gasp!) and I would frequently ask myself what would he do and how would he act when I was faced with a difficult situation. I also resisted the temptation to recreate the goofy behaviour of BC1. While I was not always successful in this resistance, at least I had a very practical (negative) target of avoidance to operate against.

My next promotion was to the Assistant Chief in charge of (as we called it in those days) the Fire Fighting Division. As the Assistant Chief, I had overall administrative responsibility for managing incident operations. I was now the boss of the Battalion Chiefs. While I had struggled throughout my career trying to learn, understand and apply some kind of fireground management system from the position of being an active participant, I now had to manage and lead a group of officers that, at that time, had a lot more 'righties' than 'lefties'. In my new position, I felt like the dog that chased the car, caught it, and now wondered what in the world to do with it (most dogs can't drive). My right-turn/left-turn experience as a company officer provided an absolutely invaluable set of real world experiences and lessons on how critical the personal performance and behaviour of the incident commander is to effective operations and to positive outcomes. As a Battalion Chief, I could also see that during active, fast moving incidents, it was virtually impossible for the incident commander to 'call the plays' quickly or completely enough to provide consistently adequate direction regardless of how personally effective that incident commander was. Effective operations require an organisation-wide plan that provides the Standard Operating Procedures (SOPs) for how the team will arrange and direct itself and the details of how specific operations will be conducted.

Along the way I also developed a healthy dose of sympathy for the incumbent company and battalion officers who were required to some how manage difficult situations they were not trained to handle. We had failed to develop any sort of incident scene management system, that would give an incident commander a fighting chance to capture and maintain control of a group of aggressive responders, who would (and still will) instinctively and automatically, blast themselves into the hazard zone – with or without orders from the incident commander.

My new Assistant Chief adventure occurred in the mid 1970s when the initial national Incident Command System (ICS) development was beginning to occur. This system development happened at an ideal time to help solve the command and control challenges we then faced. As the new Assistant Chief, I became the team leader for a small revolutionary group who imported and implemented early Incident Command System into the Phoenix Fire Department. The beginning Incident Command System focus in the national development process was directed toward outlining the basic theory of how the system worked and a description of the various system components. The original system was designed to organise and manage large scale wildland firefighting operations. These events require huge amounts of resources from lots of different agencies and typically burn a long time. They create enormous management challenges. At that time very few fire departments had attempted to adopt the system to local tactical operations and to use it to manage mostly single family and small commercial fire situations. These types of incidents make up the majority of the fires we locals are called upon to extinguish. Bringing what was in effect a large scale command system designed to manage fires in lots of burning vegetation created the need for us to remodel and scale down the larger system to fit our routine (smaller) situations. Another challenge involved converting a fairly elaborate system to plain 'implementable' English. Our officers would patiently listen, nod their heads at the appropriate times, as we would deliver our standard Incident Command System pep talk. Then they would ask (generally me) 'That's all very nice Chief, now what do you want us to do?' I then would excuse myself, go out in the hall and ask my merry band of Incident Command System implementers 'what do we want them to do?' This standard (logical, smart) question caused us to go back to the drawing board and develop an answer to our officer's question. We knew the answer had to be simple, practical, clear and doable. We also were aware that our firefighters were attracted to functional approaches – so we developed the basic functions of command and in the ensuing 25 years have developed a basic local ('home town') organisational 'structure and game plan' that drives the basic Incident Command System using the following eight standard command functions:

1. Assumption, Confirmation and Positioning.
2. Situation Evaluation.
3. Communications.
4. Deployment.
5. Strategy and Incident Action Planning.
6. Organisation.
7. Review and Revision.
8. Continuation, Transfer and Termination.

The functions connect an organisational action plan with regular tactical performance targets. They also structure a standard set of roles, relationships and functions on the strategic (IC) level, the tactical level (sectors) and the task level (companies).

The functions create a standard response, common understanding and uniform expectations of where the incident commander will be, what the will be doing and how the incident commander will interact with other members of the response team. This has created the capability to produce a consistent, reproducible command response that has become predictable and dependable. We now have a basic management structure to customise an effective organisational response to the needs of each particular incident situation. The functions provide the basis of a practical, street-orientated work plan that is routinely used for incident escalation when we encounter larger, expanding or complex situations.

Another challenge we faced was to create some sort of program management model. Even with the command functions, it was difficult to create an effective before, during and after organisational response. We had not developed the organisation skill to develop, implement, refine and make permanent the command functions. We kept making the same mistakes, going back and starting over and stumbling into each other. Simply, it was difficult to make organisational improvements 'stick'. Over a hit and miss period of a couple of years, we produced the following simple flow chart (Figure 4.1) that is still in use and serving us very well. Application of the model over time has created the capability to continually 'fix ourselves'.

Figure 4.1 Incident Command Development Model

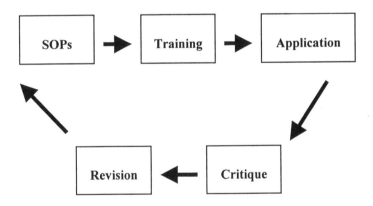

The components of the model have the following basic effect:

SOPs

Standard Operating Procedures (SOPs) require the organisation to describe standard action in a specific situation in plain, understandable language. Enough detail must be developed to produce a standard, predictable, dependable organisation-wide reaction. The SOPs create an effective starting point to get the team to all go in the same direction with common agreed upon objectives. Creating a standard starting point is a big deal during the challenging initial period of active incidents. Wherever we can create a good beginning we stand a lot better chance to create a good ending. SOPs create the capability for department members to learn and understand how the organisation will operate and such SOPs must be written to remove the mystery from the operations and activities they describe.

Training

This is where department members learn what is expected of them in a safe environment. We play the way we practise because we are able to develop a set of positive performance habits in training during training activities. It is ineffective, frustrating and dumb for us to expect our department members to do things they are not trained to do.

Application

Our customers call us when they have a problem they cannot solve themselves. This is show time for us and why we are in business. The application process is where the SOPs and training come together to mobilise the team. All of our management and operational systems must have a high capability to be actually applied to real situations. As local responders we must be consistently quick, effective and nice.

Critique

The review and critique is the standard place in our system where we evaluate how well our people and our procedures actually performed. The critique is also where we both evaluate and reinforce the content of the SOPs and what we learned in training. A standard critique should examine how effectively we connected conditions/action/outcome to produce a list of lessons learned and reinforced. The overall objective of the incident critique is for us to translate those lessons into an action plan for improvement. Such critiques should become the process and place where we evaluate how our action worked (or didn't) on a particular tactical situation. The critique becomes critical to us because it evaluates how our local resources actually performed on a local incident. While there is a lot we can learn from everyone/everywhere, it must all come together for us when we are delivering service in our hometown. The experiences and the lessons that come with that experience must be 'packaged' to be transferred throughout the organisation – this becomes a major way that we continually refine our command capability and improve our performance.

Revision

A major system management challenge is to somehow keep our operational and command system updated to match current local conditions. Things change very quickly and if an active system of surveillance, evaluation and revision is not in place, we are surprised by things we didn't know about or expect to be present. Effective revision requires every level of the organisation to actively engage what they encounter to see what has and is changing. The system must also streamline how that information is exchanged between organisational levels. The current rate of critical change requires empowered department members to not only look for change, but to also develop organisational resources to those new, unusual, sometimes surprising local dynamics. Organisational success is directly connected to developing the internal agility required to identify and take advantage of opportunities (that many times are disguised as problems). The Incident Command System becomes the 'host' for how we will actually respond and react operationally to those new, different things.

Conclusion

For the past 25 years the functions of command have created an effective foundation for our local ICS. They provide the Incident Commander with a straightforward, well known job description. Performing the functions from 1–8 leads the entire response team through a complete management and support system and provides a very practical answer to the question 'what am I supposed to do as the incident commander?' Our consistent experience has been that our incidents receive an effective command response when we can deliver a responder who initiates and continues those standard functions from the beginning of our arrival. Now, thankfully, it doesn't make any difference which direction you turn when you leave the station.

Lessons Learned

- We play the way we practise so develop positive performance habits.
- Standard Operating Procedures create an effective starting point for the team therefore whenever we can create a good beginning, we stand a lot better chance of creating a good ending.
- Critique is essential in evaluating how we actually performed, and in developing an action plan for improvement.
- Showtime for us is when our customers have a problem they cannot solve themselves: we must be quick, effective and nice!

Part II
Case Studies of Command

5 Police Commander – The Notting Hill Riot

Tony Moore
Disaster Management Centre,
Cranfield University

Tony Moore served in the Metropolitan Police for 28 years rising to the rank of Chief Superintendent. With the exception of eighteen months at New Scotland Yard as an inspector, his early career was spent on operational duties in the East End of London. As a Superintendent, he was seconded to The Police Staff College, Bramshill, for two years after which he returned to London's West End. He was one of two officers who controlled the forward incident room at the six-day siege of the Iranian Embassy in 1980. His last operational command was at Notting Hill. He was subsequently seconded to the Police Staff College, for a second time, where he was responsible for developing and running courses for Assistant Chief Constables in the management of public disorder. Throughout his police career he was involved in the policing of a number of outbreaks of disorder.

On leaving the police he took a research degree into public order policing at the University of Southampton and he is currently a senior consultant and Visiting Fellow at the Disaster Management Centre, Cranfield University, specialising in crisis management. At the time of writing, he is responsible for the design and implementation of a National Disaster Prevention and Management System in Lithuania prior to the country's admission to the North Atlantic Treaty Organisation and the European Union.

www.rmcs.cranfield.ac.uk/dmc

Introduction

Thrown with great force, the two lumps of tarmac came hurtling through the front windscreen of the unmarked police car. Oh dear! What had I got myself into? Why hadn't I gone straight home from the Metropolitan Police Training School at the end of the normal working day, instead of returning to Notting Hill Police Station?

It was Tuesday, 20 April 1982. I had just attended the second day of a course set up following a recommendation in the report by Lord Scarman (1981) that:

> the analysis of the disorder in Brixton ... and the experience in the handling of disorder elsewhere, underline in particular the need for increased training of officers, both at junior and command levels, in the handling of disorder. (p. 97, para 3.72)

For the first two days we had been at the Metropolitan Police Training School, listening to lectures and taking part in tabletop exercises. One of the scenarios had been based on Notting Hill but, as the Divisional Commander[1], I had not been allowed to take part. Instead, I had taken part in one based in Dalston, a small area of Northeast London. The following day, Wednesday, we were due to attend the Metropolitan Police Public Order Training Centre at Hounslow.

Notting Hill

Notting Hill was part of 'B' District of the Metropolitan Police at that time.[2] A small area of West London sometimes referred to as North Kensington; it had been synonymous with racial conflict and disorder, to a greater or lesser extent, ever since four days of rioting in 1958. Dubbed a 'race riot', the disorder on that occasion was predominantly between black and white youths, with the police in the middle trying to keep the peace. In 1976 there had been serious disorder on the second day of the annual Notting Hill Carnival, when a section of the crowd threw bricks, bottles and other missiles at police officers, whose only defence to this violent attack was dustbin lids, empty milk crates and plastic 'no parking' cones. Over 400 police officers were injured. Following this, plastic protective shields were introduced. In 1980, serious rioting had broken out in Bristol following a raid by the police on the Black and White Cafe in the St Paul's district, an area not dissimilar to the All Saints Road area of Notting Hill. In 1981, in April, there were three days of rioting in Brixton in south London. Immediately following the riots, Lord

Scarman was appointed to investigate the causes, but, before he could complete his report, there was more rioting in July, in a number of towns and cities throughout England, most notably in Birmingham, Liverpool – where CS gas was used for the first and, to-date, only time against a crowd on the British mainland,[3] and one person was killed when he was run over by a police vehicle – and Manchester. There was minor disorder in Notting Hill on this occasion but it was nothing like on the scale that occurred in other areas.

The problem area in Notting Hill in 1982, as it had been for a number of years, was centred on All Saints Road. It was, and indeed still is, a fairly narrow street running from Tavistock Road in the north to Westbourne Park Road in the south. Crossing it were two highways, St Luke's Mews and Lancaster Road, both of which run from east to west. Also running from east to west from the east-side of All Saints Road is McGregor Road (see figure 5.1).

Figure 5.1 Notting Hill

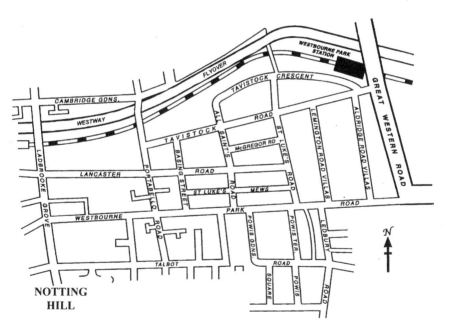

With the exception of the Apollo Public House on the corner of All Saints Road and Lancaster Road, that part of All Saints Road which runs between Lancaster Road and Westbourne Park Road consisted of terraced three-storey buildings, all abutting onto the pavement. The buildings were, in the main, used as small shops, cafes, etc., at ground floor and basement level with the upper two floors being used for residential purposes.

Whilst the majority of the population went about their lawful businesses, a minority, not all of them resident in the Notting Hill area, were engaged in the buying and selling of cannabis and other drugs, running illegal drinking dens, known as shebeens, and some were involved in street robberies and burglaries.

First Incident – a Bottle is Thrown

At about 7 p.m. on 20 April 1982, a bottle was thrown at two young police officers as they patrolled All Saints Road on foot. The bottle, which was thrown from behind the officers, just missed one of them before smashing into the wall. Unable to identify the thrower, or, indeed, the direction from which it had come, the officers continued their patrol as if nothing had happened.

Second Incident – a Prisoner is Lost

At about 9 p.m., three police officers, two male and one female, on a routine patrol stopped a man and a woman in Portobello Road, about ten yards north of the junction with Westbourne Park Road, after they had received information from another police officer that one or both of them had just purchased drugs in All Saints Road. They questioned the two people and found a piece of resinous substance in the woman's pocket. This was later analysed and found to be 27.5 grams of cannabis resin, which, at that time, would have fetched about £70 on the street market. The woman was arrested and one of the officers radioed for a police van to convey them to the police station. Nothing having been found in the man's possession he was told he could go. At that point, a number of people, both male and female, converged on the officers from a local fast food take-away in Portobello Road and released the woman from custody. All the people, including the woman, ran to All Saints Road. Two of the three police officers were injured in this confrontation.

A number of police area cars[4] responded to the radio message from one of the three officers requesting urgent assistance. By now there was a large, hostile crowd in All Saints Road and two of the responding police cars were damaged by missiles thrown at the vehicles as they were driven westwards along Westbourne Park Road. As a result, one of the cars sent the following message to the Metropolitan Police Control Room at New Scotland Yard:

> Information for all units. If you go past this road they're going to stone us. Suggest that the only way to deal with this problem is for shields. Do we have a Senior Officer to authorise it?

The Riot

I was conscious of the fact that, in his report published in November 1981, Scarman had said that 'the disorders [in Brixton in April 1981] revealed weaknesses in the capacity of the police to respond sufficiently firmly to violence in the streets' and, that 'officers untrained in the command of men carrying [shields] found themselves thrust into the front line.' (p. 71, paragraph 4.90). Various experiences gained as I came up through the ranks convinced me that it was essential for police commanders to be both familiar and comfortable with the tactics being taught to the officers they were likely to command in riot situations. This, together with the fact that I was aware of the history of disorder and racial conflict in Notting Hill, led me to ensure that I was on the first of the three-day courses for officers of my rank being run at the Metropolitan Police Training School.

I had returned to Notting Hill Police Station that evening because I knew that six members of the Joint Services Staff Course, a prestigious course attended by senior military officers from all over the world,[5] were visiting B District that night and two of them would spend time on Notting Hill Division. It was my intention to be present to brief and debrief them. In the intervening period they would spend some time talking to police officers in the various departments at the station and patrolling the division in a police car.

I was sitting in my office at Notting Hill Police Station waiting for the two military personnel to return to the station following a period in a patrol car, when the above message came through at about 9.20 p.m. I was dressed in civilian clothes.

I did not want police units to respond to this incident in a haphazard fashion as had so often been the case in the past in London and elsewhere. An operational plan existed for responding to serious incidents in All Saints Road and I immediately gave instructions for a message to be sent over the radio directing units to stand-by positions in accordance with the plan. The standby positions were deliberately located sufficiently far away from any likely flash point as not to make the crowd aware of their deployment. The fact that, on this occasion, units were deployed to their standby positions, without further prompting, by a constable who was driving one of the patrol cars at the time, showed the value of such plans being widely known and regularly discussed. As a result, police units were deployed to locations as follows:[6]

- Lancaster Road junction with Ladbroke Grove B30, E30 and F30.
- Westbourne Park Road junction with Ladbroke Grove H30.
- Ledbury Road junction with Westbourne Park Road F Zero and 2 R/T Cars.
- Tavistock Road junction with St Luke's Road D30 and DR22.

I also gave an instruction that all police units converging on the area were to do so silently. On a number of previous occasions in both Notting Hill and elsewhere I had experienced situations in which large numbers of people had been drawn to locations by the sight of police vehicles, using blue flashing lights and sounding their audible warning instruments, going to a particular area or location and I did not want this to happen on this occasion.

By now I had changed into uniform and, at about 9.25 p.m., I was driven from the Police Station by Sergeant Hole. I was sitting in the front passenger seat. Behind me, in a rear seat, sat Inspector Graham Sharpe, the duty inspector. My intention on leaving the station was twofold:

- to visit the scene to assess the situation, and
- to search out one of the local 'street leaders' to discuss with him how we could, together, defuse the situation.

We approached All Saints Road by driving in an easterly direction along Westbourne Park Road. This was a mistake because when we reached a point approximately 30 yards west of All Saints Road we were suddenly confronted by a crowd of about 50 people who immediately started attacking the car with missiles. had been too pre-occupied thinking about how I was going to deal with the situation and what I was going to say to any of the 'street leaders' to notice the route Sergeant Hole was taking. Because the events of that evening were totally unexpected, my normal driver was not available. I had visited All Saints Road on a number of previous occasions when tension was high. From experience, I knew that when trouble did erupt, it did so in the southern half of All Saints Road, between Lancaster Road and Westbourne Park Road. On these occasions, my regular driver would invariably skirt the area and come in from the northern end of All Saints Road, i.e. from Tavistock Road.

Two fairly large pieces of tarmac crashed through the front windscreen of the car. One hit Sergeant Hole on the shoulder; the other flew between us and landed on the back seat next to Inspector Sharpe. Despite being slightly injured, Sergeant Hole immediately threw the car into reverse and we speedily backed down as far as Portobello Road. At the same time I ordered all the Immediate Response Units from their stand-by positions into All Saints Road, with instructions to disperse the crowd. B30, E30, F30, D30 and DR22 entered All Saints Road from the Lancaster Road end. Some missiles were thrown and DR22 suffered a broken windscreen but the crowd rapidly melted away on the arrival of the police vehicles. Many of them ran into the terraced houses on either side of the street. One person was arrested.

The Second Withdrawal

I was mindful of the sensitivity of any police operation in a mainly ethnic area at that time. The Scarman report (1981, p. 64), had criticised the police for 'errors of judgement' and 'a lack of imagination and flexibility' when policing such areas[7] and I was anxious not to draw unnecessary criticism on the Metropolitan Police Force. Hearing that the crowd had been dispersed, at 9.32 p.m. I ordered all the police units to leave All Saints Road and return to Notting Hill Police Station. My purpose in doing this was two-fold:

- to, hopefully, allow the situation to cool down.
- at the same time, I wanted to re-group my resources and be in a position to redeploy in a more orderly fashion than had been the case previously.

However, I was conscious of the judge's comments about the police withdrawal from St Paul's, a mainly ethnic area of Bristol, in 1980. On this occasion, the police had been overwhelmed following a raid on the Black and White Cafe and the local Chief Constable took the decision 'to withdraw to regroup, to gather strength and to obtain sufficient reinforcements to ensure a speedy return to law and order with a minimum of bloodshed.' In his report to the Home Secretary, the Chief Constable said, 'it was hoped that the removal of police – the object of the violence – would quieten the crowd and itself help the return to order'.[8] However, in the subsequent trial of some of the rioters, the judge called it 'a period of re-arming, not a prelude to normality.' (see Harris, Wallace & Booth, 1983).

I was also aware of the fact that, in their evidence to the Scarman Inquiry, mediators had criticised the police for not withdrawing when requested to do so at the height of the Brixton riot in April 1981, on the grounds that 'since the fury of the crowd was directed at the police, that fury would cease if its object was removed.' Scarman (1981, p. 70) rejected this criticism on the grounds that 'arson and looting in Railton Road were already under-way by the time of the attempted mediation' and there is little doubt that Commander Fairbairn[9] would have been heavily criticised had he done so. However, the situation on this occasion was slightly different. There were no mediators immediately to hand and, at the time of the withdrawal I was unaware of any arson or looting.

At the police station I realised that time was short if I had to gain the initiative as quickly as possible before any serious damage was done. I had a clear idea of what I wanted to do if it became necessary; consequently I gave the officers-in-charge of all the units – some were inspectors and some were sergeants – a brief outline of what had occurred and said that I hoped the situation would defuse itself after the dispersal of the youths. However, in case it did not, they were to prepare themselves for a more precise operational deployment, which meant getting dressed in flame-retardant overalls and wearing protective helmets. They were also to be ready to equip themselves with the new, small, round protective shields introduced into the Metropolitan Police following the serious shortcomings in both equipment and tactics at Brixton the previous April. Scarman (1981 p. 97) had suggested that the use of the long shields at Brixton had encouraged 'officers to adopt a largely defensive posture' which served 'to attract missiles from a crowd' with the result 'that lines of police officers behind the shields effectively become 'Aunt Sallies' for the crowd to aim at.' I was determined that this would not be the case. There was no discussion, merely an opportunity at the end to ask questions to clarify the directions I had given.

Re-deployment

Between 9.45 p.m. and 10.10 p.m., fourteen telephone calls were received in the Metropolitan Police Control Room at New Scotland Yard from members of the public which suggested that, in the vicinity of All Saints Road:

- youths were arming themselves with bricks;
- barricades were being erected;
- petrol bombs were being manufactured, using bottles, cloth wicks and petrol.

One caller estimated the number of people in the street to be about 500; others estimated it at slightly less. Whilst I had no confirmation from police sources that the information I was receiving was correct, and whilst there have been occasions when misleading information has deliberately been fed into the police by individuals and groups for personal motives, I had no reason to doubt the general accuracy of the information on this occasion because I was aware that some of the calls came from residents who overlooked the area but were not part of the group that normally frequented All Saints Road. In fact, the information turned out to be remarkably accurate.

As these telephone calls were being received and passed to me over the radio, I deployed units again to stand-by positions as follows:

- Lancaster Road, 50 yards west of Ladbroke Grove. B30, F30.
- Westbourne Park Road, 50 yards west of Ladbroke Grove C30, G30, G31.
- Great Western Road junction with Tavistock Road D30, H30.
- Ledbury Road about 50 yards south of Westbourne Park Road E30, BH2, N30.
- Powis Square (south side) No 6 Unit, Special Patrol Group. [10]

By 10 p.m., I had taken up a position in a police car at the northern-most end of All Saints Road – just to the north of the junction with Tavistock Road. I had a map of the area on my lap; a police constable was acting as my driver and another constable as my radio operator.

As more information came to me from telephone calls from members of the public via the Central Control Room at New Scotland Yard, at 10.05 p.m., I moved some of the units closer to the scene. B30 and F30 were instructed to move forward in Lancaster Road to a position approximately 50 yards west of Portobello Road. The standby position at their original location was filled by No. 8 Unit of the Special Patrol Group. C30, G30 and G31 were instructed to move forward along Westbourne Park Road to a position approximately 50 yards west of Portobello Road. Their original standby location was filled by No. 5 Unit of the Special Patrol Group.

At 10.05 p.m., I gave instructions that, on my word of command, units were to move from their standby positions to the following locations:

- All Saints Road junction with Lancaster Road H30, D30, B30, F30.
- All Saints Road junction with Westbourne Park Road No 6 Unit Special Patrol Group, E30, G30, G31.
- St Luke's Mews junction with Basing Street C30.
- St Luke's Mews junction with St Luke's Road BH22.

At about 10.10 p.m., as a result of the information contained in the telephone calls being received from members of the public by the Central Control Room at New Scotland Yard, I was able to tell units that there appeared to be barricades across:

- All Saints Road at its junction with Westbourne Park Road.
- St Luke's Mews (both sides) at the junction with All Saints Road.
- Lancaster Road (west side) at its junction with All Saints Road.

At 10.12 p.m., I gave my final instructions over the radio:

> For the information of all units on this operation – when I give the word to move in, I want you to move in and I want as many arrests as possible.

The message was repeated by the controller at Force headquarters.

At 10.14 p.m., I gave my order:

> I want all units, all units to move in now – all units to move in now, demolish the barricades, as many arrests as possible.

With that, all the units moved in quickly and silently, taking the crowd in All Saints Road completely by surprise. At All Saints Road junction with Westbourne Park Road, a British Leyland 1100 had been placed at right angles across the street and doused in petrol. In St Luke's Mews (west) a van had been turned on its side; various obstacles such as old doors and other pieces of wood and galvanised steel had been placed around it to make a barricade. In St Luke's Mews (east) a Rover motor car had been parked at right angles across the mews. In Lancaster Road, about 20 yards west of All Saints Road, a builders skip and a British Leyland 1100 car had been placed at right angles across the road.

Twenty-five petrol bombs, primed and ready to light, were seized by the police, together with cans and plastic containers containing further supplies of petrol, at two different locations. One cache of petrol bombs was found at the junction in All Saints Road at the junction with Lancaster Road; a second cache was found in All Saints Road at the junction of St Luke's Mews. The bottles of a third cache, which was in All Saints Road at the junction with Westbourne Park Road, were quickly broken as the police arrived.

Twenty-six arrests were made at the time, which together with the person arrested at about 9.30 p.m., made 27 in total.[11] They were charged with a variety of offences including the possession of offensive weapons (petrol bombs and bricks), threatening behaviour, assault on police and using insulting or threatening words.   It was relevant to note that only 3 of those arrested resided in the vicinity of All Saints Road and only 8 on Notting Hill police division.

I remained where I was, hidden from view, until at 10.19 p.m., 5 minutes after I gave the order to move against the barricades, when one of the Immediate Response Units reported the road to be all clear.   Despite my closeness to the scene, I did not want to become embroiled in any running battle that might develop.   It was essential that I remained apart from what was going on as the police entered All Saints Road in case further deployments were necessary.

Criticisms of my Actions

Amongst the criticisms subsequently levelled at me were four, which have a particular significance.   Firstly, why did I go to the scene, particularly on the second occasion?   Why did I not stay at Notting Hill Police Station where maps and communications were abundant?[12]   It must be remembered that this incident occurred before what is commonly known as the Gold, Silver, Bronze concept of command (see Arbuthnot this volume) was introduced.[13]   Until just before I deployed for the second time, when a chief inspector, who had no experience of public order policing, arrived from another division, I was the only officer above the rank of inspector immediately available.   Additionally, I wanted to be at the scene immediately following the deployment of the units to ensure that only such force as was necessary to disperse the crowd and restore order was used.

Secondly, following on from this, there was criticism of my instruction that I wanted as many arrests as possible.   In a planned demonstration, it is possible to give detailed briefings to those who are to be deployed in the event of disorder.   In this case it was not.   The various units had come from all over London in response to a worsening situation and there had been insufficient time to give them a detailed briefing.   By focusing the minds of the officers on arrests I hoped to reduce the possibility that the amount of force used in the dispersal of the crowd would be excessive.

In addition, as soon as I was aware that arrests had been made, I sent an instruction by radio to Notting Hill Police Station that a doctor should be summoned to examine everyone who had been arrested.   Twenty-four of those arrested were subsequently seen by a doctor at the police station.   A number had superficial cuts, etc., consistent with struggling whilst being arrested, and one was taken to hospital where he was detained for observation because he claimed he had been unconscious for a short while, although no apparent injuries were found. He was released from hospital the following day. Three of those arrested refused to allow the doctor to examine them other than visually. There was not a single complaint that excessive force had been used.

Thirdly, why did I box the rioters in? By having units approaching from every direction, I left the crowd with no escape route. In his report into the Red Lion Square riot of 1974.[14]   Lord Scarman (1975) had criticised the police for not leaving an escape route when they were deployed to disperse the crowd, saying 'it is important for the officer in charge to ensure, wherever possible, that a crowd has sufficient means of moving away before taking action to disperse or disrupt it.' (p. 40) I took the view that the circumstances in this case were different.   At Red Lion Square it was a political demonstration and there can be no doubt that many peaceful protestors were caught up in the crowd that had been boxed in by the police. At Notting Hill, however, any people who were of peaceful persuasion had an opportunity to leave the area following the withdrawal of police units on the first occasion they entered All Saints Road. Anyone seeing the activity that followed, the barricades being erected and the arming of people with bricks and petrol bombs, should have been aware that the police were likely to take some positive action and therefore had an opportunity to leave the area right up to the time the police were ordered in.

Fourthly, the crowd was given no warning of the impending police action. In his report into the Red Lion Square riot, Scarman (1975) claimed that warnings should generally be given because it enables those 'without violent intent' to leave and those who remain cannot claim 'that police action came as a total surprise'. (p. 40) Advice given to police commanders today suggests that before taking any overt action, such as a baton charge, a warning should be given to the crowd, but, in his report into Red Lion Square, Scarman indicated that 'there will be some occasions where the police need to keep the element of surprise in order to secure the success of their operation.' (p. 40) I was concerned to ensure that there would be the minimum number of injuries both to my own officers as well as the public. I am convinced that, had such a warning been given to the crowd, the barriers erected across the street would have been set alight and the advancing officers would have been in considerable danger from the petrol bombs that would almost certainly have been thrown at them.

Conclusions

Attempts to defuse the situation by twice withdrawing having failed, I believe I was left with no alternative other than to bring the matter to a swift conclusion at about 10.15 p.m. To have delayed the efforts to restore order, even for five minutes, would have enabled the crowd to have set light to at least one of the barricades and there is little doubt they would have been ready to throw the lighted petrol bombs at the advancing police officers, thus escalating the situation considerably.

It is generally recognised that stress is a natural reaction to unnatural or unusual events. I was not conscious of suffering from an increased adrenaline flow as the events unfolded but, rightly or wrongly, there was a sense of euphoria later that evening and on the days following the 20 April.

After all the criticisms of the police response to the disorders of 1981 in various parts of the country, the units that responded to the disorder in Notting Hill had done so extremely effectively showing the benefits of the training they had undergone in the previous nine months. This was generally reflected in the media and in parliament.

The following day, the events shared the front page of the Sun newspaper with a report that the British Task Force had retaken South Georgia from the Argentineans; the Daily Express described it as a 'battle'. The Standard reported, in huge headlines across its front page, 'Notting Hill Riot – 24 held'; and the Kensington News and Post headlined its front page with 'All Saints riot flares.' The following weekend, The Sunday Times analysed the incident in a fairly lengthy article.[15]

Six days later, on 26 April, in a statement to the House of Commons, the Home Secretary, William Whitelaw, described it as 'an example of the sort of action which has to be taken quickly and decisively'.[16]

Reflections

Not surprisingly I had never heard of Gary Klein or of Naturalistic Decision-Making in 1982. As a police officer I had attended courses in what was originally called man-management and which subsequently became known as management training as I rose through the ranks but, with the exception of the three-day course I was attending at the time of this incident, none of it related to decision-making when faced with a rapid onset chain of events. Since then, however, Gary Klein (1998) and others have emphasised the importance of experience when making decisions in rapidly developing, unstructured situations. Shortly after this riot, I was posted to the staff at the Police Staff

College Bramshill where I became familiar with a book by an eminent military historian, Liddell Hart (1967) called *Strategy: The Indirect Approach,* In it, Liddell Hart argues that there are two kinds of experience, direct and indirect.

Direct experience is self-explanatory. In the context of this chapter, it is actually responding to a riot. My direct experience had involved the policing of a number of disorderly situations. These included the St Pancras Rent riots in 1960 as a constable, the anti-American demonstrations over Vietnam in and around Grosvenor Square in 1968 as an inspector, and at least three incidents in East London, between 1972 and 1976, as a chief inspector. Additionally during this latter period I had spent four years policing Arsenal football ground on a regular basis, at a time when so-called football hooligans regularly went on the rampage. Between 1979 and 1982 I had policed a number of demonstrations in Central London, some of which involved clashes between right and left wing movements, e.g. the National Front and British Movement in conflict with the Anti-Nazi League, and the annual Notting Hill Carnival. Despite this, the number of occasions on which individual police commanders are faced with a spontaneous outbreak of disorder in contrast to disorder which occurs at a pre-planned event, e.g. the Poll Tax riot in Trafalgar Square in 1990 and the City of London riot in 1999, is rare.

Indirect experience, on the other hand, is achieved by training and studying history amongst other things. I had attended the Metropolitan Police Public Order Training Centre at Hounslow on a regular basis and had studied past incidents of disorder, learning lessons from both positive and negative actions. Unlike military commanders who seem to put pen to paper as soon as a battle is over, police commanders have generally appeared extremely reluctant to do so. Consequently, there are very few personal accounts of the response to disorder. However, following the inner-city riots of 1980, 1981 and 1985 and the industrial disorders associated with the miners' strike in 1984/1985 and the printers' strike in 1986/1987, a number of books on such events were written by, in the main, academics. This trend continued into the 1990s. But, prior to the events in Notting Hill in 1982, there were only a few accounts available. In addition to those mentioned earlier, the following are worthy of note: (Constable, 1970; Critchley; 1970; Dummett, 1980; Kerner, 1968, Mather, 1959; McCone, 1965; Thurston, 1967; Williams, 1967).

As I saw the inadequacies of the police in 1980 and 1981 in responding to serious disorder, and, in particular, the inability of some police commanders to comprehend what was required, I had given a great deal of thought to what I would do if and when I was faced with such problems.

A number of people have since written on this subject. For instance, writing nearly ten years after the events of the early eighties, Waddington

(1991) suggested that the police continually failed,

> to appreciate the nature of the task of quelling serious disorder. They, and many others, remain blinded by their traditional image of policing public order without recourse to overtly aggressive tactics. From the vantage-point of 1990, the development of police public-order tactics is confused and out of touch with reality. (p. 159)

In relation to the 1981 Brixton riots, former Home Secretary William Whitelaw (1990) subsequently wrote in his autobiography, 'As a trained soldier, I was struck by the immobility of the police response on the Brixton streets'. He went on to suggest that the police should consider 'outflanking movements, so much a part of military folklore' and that the rioters should be 'harried and kept on the move'. (p. 244) And, in an article written shortly after I had spent four years running courses in the management of public disorder at the Police Staff College (Moore, 1986, p. 89), I suggested that it was,

> a fallacy to believe that every senior or middle-ranking police officer will make a good incident or ground commander. In the same way as there are those who have a talent for computer or communication systems, administration, organisational planning (as opposed to operational planning) or criminal investigation, there are those who have a similar talent for the handling of public disorder. The most capable officers in running a division on a day-to-day basis are not necessarily those who will stand up best either to a spontaneous eruption of disorder or indeed to the strain of prolonged public disorder.

It follows, therefore, that:

> the selection and training of police commanders to deal with such eventualities is crucial.

When I made my first foray into All Saints Road, I was the only senior officer present at Notting Hill at the time. Nevertheless, it could be argued that I should not have placed myself in a position where I could have been seriously injured. On previous occasions, the frequenters of All Saints Road were fairly respectful of rank and I never found myself on the receiving end of the kind of abuse to which constables, sergeants and even inspectors were subjected. Added to that, my deputy, Superintendent Gwen Symonds, had only ten days previously responded to an incident in which a crowd in All Saints Road had started to erect barricades. On this occasion she had been able to contact at least one of the 'street leaders' and to defuse the situation. Additionally, there had been occasions in the past when I had negotiated a peaceful solution to situations, which had the potential for serious conflict. I had, therefore, not considered the possibility of me being attacked. In other words, I had not gone sufficiently down the 'what if?' path.

This first foray into All Saints Road was disjointed and disorganised with no overall plan as to what to do once the units got into the street, other than disperse the crowd. Compared with that, the second foray was highly organised. As information built up that the street was not going to quieten down, and that barricades were being put in place and petrol bombs prepared, I had a brief amount of time to choose an option, given my knowledge of public order policing in general, coupled with my knowledge of the basic plan, the area and the people who frequented it.

- Mediation is always the first option. But history has shown that this often fails and it is important to be in a position to implement alternative strategies.
- Any plan to deal with disorderly crowds should be kept as simple as possible.
- If the initiative has been lost to the rioters during the early stages, it must be recovered by the police as quickly as possible.

Lessons

There are three essential elements for success in dealing with any public disorder situation:

- *Proper launching, which includes planning, the provision of equipment and the build-up of resources, particularly if the outbreak of disorder occurs suddenly.* If the deployment of police units is planned on sound strategic and tactical understandings of the principles of dealing with public disorder, the police are more likely to gain the upper hand quickly.

- *The courage, initiative and skill of police commanders once disorder has broken out.* In order to respond effectively, police commanders, in their various grades, must have a sound understanding of the techniques of staging the many and varied operations that their officers may be required to undertake in public disorder situations, whether defensive or offensive. The aquisition of such an understanding requires both, a study of history and rigorous training.

- *The professionalism, confidence and morale of the officers who go to make up the various units responding to the disorder.* Once again, rigorous training is vital if increased professionalism is to be achieved. It is essential that police commanders involve themselves in all aspects of planning and training so that the officers who subsequently respond will have confidence in their ability to restore order both effectively and efficiently.

If any of these three is missing the situation may well be lost.

Notes

1. At that time, a Divisional Commander in the Metropolitan Police held the rank of Chief Superintendent. Notting Hill Division was staffed by approximately 270 sworn officers and 50 civilians.

2. Two other divisions, Chelsea and Kensington, made up the District. The District headquarters were attached to Kensington Police Station.

3. The use of CS gas on this occasion was subsequently severely criticised in some quarters. Because the police were in danger of being over-run, the Chief Constable authorised the use of a type of CS gas which was designed to penetrate barricades when hostages were being held by criminals or terrorists. It had not been designed for use in crowd situations.

4. An area car normally had a crew of two, a driver and a radio operator.

5. Two senior police officers, one from the UK and one from overseas, also attended the course.

6. At that time, an Immediate Response Unit normally consisted of 1 sergeant and 10 constables in a protected vehicle. Two Immediate Response Units would normally be under the command of an inspector. All Immediate Response Units used the call signs 30 or 31 preceded by their District letter, and all officers had protected helmets and flame-retardant overalls. Long shields and the new small round shields were carried on each vehicle. F Zero was a dog unit containing two constables and two dogs. DR22 was an unprotected vehicle with an ad-hoc crew.

7. Whilst I was not conscious of thinking specifically about the Scarman Report during this time, in the five months between publication of the report and the date of these events, i.e. 20 April 1982, I had given a great deal of thought to what I would do if faced with serious disorder in Notting Hill. Additionally, I had discussed the subject with senior staff, i.e. the superintendent, chief inspectors and inspectors at the normal monthly management meetings.

8. The Report to the Home Secretary by the Chief Constable of Avon & Somerset, as reported in *Police, the Journal of the Police Federation of England and Wales,* May 1980, p. 11.

9.   Commander Fairburn was the officer in overall command on the ground during the worst of the rioting in Brixton on 11 April 1980. In today's terms, he would have been the Silver Commander.

10.   A Special Patrol Group Unit then consisted of an inspector, 3 sergeants and 30 constables with appropriate transport.

11.   A further 10 people were arrested during the following days.

12.   I later learned that, as soon as the events began to unravel in Notting Hill, the Central Control Room inspector at New Scotland Yard had instructed a sergeant to sit alongside the constable who was already operating the dedicated radio channel. The sergeant plotted the deployments I was making on a map so, although I did not know it at the time, there was a back-up record of the deployments I was making on the map, which I held on my lap.

13.   The three-tiered system of command currently in use in Great Britain was first introduced by the Metropolitan Police in 1986 following the review of the serious riots, which occurred in Brixton and Tottenham in the autumn of 1985.

14.   The riot in Red Lion Square arose when various left-wing groups opposed a march by the National Front. During clashes between the police and the left-wing groups, Kevin Gately, a student at Warwick University, became the first demonstrator to die during a riot on the British mainland for over fifty years.

15.   The Sun, 21 April 1982; The Express, 21 April 1982; The Standard, 21 April 1982; The Kensington News and Post, 23 April 1982; The Sunday Times, 25 April 1982.

16.   Hansard, 26 April 1982, p. 107, col. 1.

References

Critchley, T. (1970) *The Conquest of Violence*, London: Constable.
Dummet, M. (1980) The Report of the Unofficial Committee of Enquiry, National Council for Civil Liberties, London.
Harris, J., Wallace, T. & Booth, H. (1983) *To Ride the Storm: The 1980 Bristol Riot and the State*. London: Heinemann.
Kerner, O. (1968) *Report of the National Advisory Commission on Civil Disorders*, New York: Bantom Books.
Klein, G. (1998) *Sources of Power: How People Make Decisions*. Cambridge: Mass: MIT Press.
Liddell Hart, B. (1967) *Strategy: The Indirect Approach*. London: Faber & Faber.
McCone, J. (1965) *Violence in the City – an end or a beginning? A Report by the Governor's Commission on the Los Angeles riots*, Los Angeles.
Mather, F.C. (1959) *Public Order in the Age of the Charterists*, Manchester University Press.
Moore, T. (1986) Public order, The police commander's role. *Policing*, 2,2, Summer, 88–100.
Scarman, Rt. Hon. Lord Justice (1975) *The Red Lion Square Disorders of 15 June 1974: Report of an Inquiry* (Cmnd 5919) London: Her Majesty's Stationery Office.

Scarman, Rt. Hon. Lord Justice (1981) *The Brixton disorders 10 – 12 April 1981: Report of an Inquiry* (Cmnd 8427) London: Her Majesty's Stationery Office (HMSO).

Thurston, G. (1967) *The Clerkenwell Riot: The Killing of Constable Culley*, London: George Allen & Unwin.

Waddington, P. (1991) *The Strong Arm of the Law*, Oxford: Oxford University Press.

Whitelaw, W. (1990) *The Whitelaw Memoirs*, London: Headline.

Williams, D. (1967) *Keeping the Peace*, London: Hutchinson.

6   Fire Commander

Dennis Davis
HM Chief Inspector of Fire Services, Scotland

Dennis Davis became a firefighter in 1965 at the age of 18, joining the Walsall County Borough Fire Brigade, where he progressed to the rank of Sub-Officer. He joined Cheshire Fire Brigade as a Station Officer in 1971 and moved through the ranks of that brigade to become Deputy Chief Officer in 1983 and subsequently Chief Fire Officer in 1986. Because of the risk profile of the county of Cheshire he gained significant experience of command at major petrochemical incidents and fires.

He is a Chartered Engineer, a Life Fellow and Past President of the Institution of Fire Engineers and its current Management Committee Chairman. He has also served as President of the Chief and Assistant Chief Fire Officers' Association (CACFOA) and was for several years Chairman of that Association's Emergency Planning Committee. He was appointed to his present position in May 1999.

Introduction

It was a pleasant February evening, which had begun with dinner for six at a local restaurant. Unfortunately, as so often happens in any fire brigade commander's life, the sounding of my paging unit at just after 8.45 p.m. was to disrupt this happy scene alerting me to the fact that a local major chemical company was experiencing a fire.

Having been with the Cheshire Fire Brigade for over 20 years, in an area holding one of the largest conglomerations of petro-chemical industries in the country, such calls were no surprise. I had, after all, attended many major chemical plant incidents, and having been heavily involved in the Staff Department, had spent quite some time working with the industry pre-planning and designing chemical emergency response procedures like the one called 'Cloudburst' which was now swinging into action.

So, nothing too unusual for a firefighter about a disrupted meal or a call to a major chemical manufacturer's site. However, as the first appliances on the predetermined attendance started to respond, circumstances dictated an event that was to ultimately reach national proportions, with calls for a Public Inquiry and media representatives who scented a disaster had been avoided more by good fortune than cohesive response. It was an accepted fact of life in Cheshire that when incidents of this type occurred there would be considerable press interest. Major television and radio stations were serving around 10 million people in the adjacent area. There was also an underlying safety concern, about such petrochemical installations, amongst the public. Managing those legitimate interests was always a pressure on the Incident Commander. Such a background naturally made any fire brigade commander alert to the public consequences of toxic material escapes and to the fact that the public often had a perception of a higher risk than was the reality.

The Incident

The fire that occurred and had called for the initiation of the major response procedure referred to as Cloudburst was at the premises of Associated Octel, situated at Oil Sites Road, Ellesmere Port, Cheshire. Subsequently, although this of course was not known at the time, this incident was to become serious enough to be reviewed and appraised by the Health and Safety Executive following a number of concerns raised by residents.

The published report (HSE, 1996) of the incident which occurred on the 1st and 2nd of February 1994 provides a great deal of detail about the incident. This will not be covered in this chapter, which concentrates more upon the decision processes used by various commanders as the incident progressed.

It is, however, worth mentioning that this 87 acre site in north west Cheshire is just over 1.5 km from the populated area of the town of Ellesmere Port and borders on its northern side the Manchester Ship Canal and River Mersey. The site itself is very much a part of the Stanlow industrial complex of oil and petro-chemical manufacturing sites. The whole site is important to petroleum production since manufactured on-site are tetraethyl and tetramethyl lead used as anti-knock compounds in engine fuel. As such it affects the production of all other motor fuels, there being no other similar site in the UK. Something else for the firefighter to think about are the sodium and chlorination plants, also present on-site with their attendant risks. The company therefore operate a notified major hazard site presenting the hazard to the wider environment, as described within UK regulations derived from the European 'Seveso' Directive, which arose after dioxine escaped into the Italian countryside with devastating consequences in July 1976 (see HSE, 1999).

The Hazard

The initial call had been made as a result of a leak of ethyl chloride. This was both serious on-site and had the possibility of affecting premises and the public off-site. The wind speed and direction indicated was twelve to fourteen knots from a westerly direction. This provided sufficient information to enable a joint Emergency Services Reinforcement Base (ESRB) to be established at Ellesmere Port Fire Station.

To understand the following account of the management of the incident, it should be noted that the structure of the Cloudburst command procedure included both an on-site, Works Emergency Headquarters, and off-site emergency centre, the ESRB, to be in operation for this sort of incident. This directly impacts on the decision process that is used. The objective of having these two centres is to reduce actual attendance levels on-site, so lowering the number of persons at risk, allow briefing of fire-fighters to occur before they enter the risk area, and move logistical support away from the active operational zone. The overall Incident Commander operates from the ESRB.

The Operation

Returning to the initial incident, the fire service Station Officer on arrival established an on-site command centre at the Works Emergency Headquarters and a Forward Control point close to the actual incident. He then committed four firefighters dressed in breathing apparatus and chemical protection suits to position water ground monitors, so adding to one ground monitor already positioned by the works fire brigade, to form a water curtain. This effectively gave three water sprays that were used to disperse the gas cloud that was emanating from the spilt products.

Unfortunately, at the time, one of the key pieces of information not known to the on-site Station Officer was that the product was highly flammable. The information he was using at the time had been gained during earlier pre-planning risk assessments using inspection processes common to all UK Brigades. Information obtained from these pre-incident visits conducted under risk inspection procedures was then carried on appliances in paper form. This information was on arrival with shift staff. It suggested that the real hazard from the leaking product was a toxic one created by the evolving hydrochloric acid gas. It was this gas that he was seeking to disperse.

With these initial procedures in operation, the brigade Fire Control mobilised, in accordance with the Cloudburst procedure, further sufficient pumping appliances to the ESRB to enable an ongoing operation to be conducted at the site. This response in one way also reflected a health warning gained from previous experience since incidents of this type frequently suffer from the simple difficulty of accurate information transfer. Misinformation can often start with the company getting the wind direction wrong for the turnout message – not insignificant if you are responding to a toxic release and coming at it in the hazard direction because of a basic error. Confirmation of basic information was therefore the natural first action and in this case support was already available through the attendance of the on-site fire service.

Shortly after the fire brigade's arrival one works employee had been conveyed to hospital suffering the effects of fumes. This tended to reinforce the view that a major toxic risk existed. It therefore appeared appropriate to continue the operation as initiated with water sprays whilst decontaminating the various firefighters who became involved as the operation progressed.

On-site Communications

On-site, communication to the Works Emergency Headquarters and thence to the ESRB were set up and at the ESRB, further links to the District Off-site Emergency Centre (DOSEC) were established. The DOSEC was responsible for co-ordinating local authority responses should the incident progress and again are part of the established Cloudburst emergency plan. Having the DOSEC remotely located also ensured the ESRB was free of all but essential emergency response staff and media representation. That helped avoid not only distraction but also aided concentration on the key issue of the incident.

Escalation

The Station Officer had, by 9.15 p.m. been supported by an Assistant Divisional Officer who had taken charge of the incident and increased the pumping appliances on-site to six. This facilitated the support by additional firefighting resources in the terms of water monitors whilst deploying additional breathing apparatus wearers.

Off-site, monitoring was being conducted in accordance with standard procedures and adjacent industrial sites had been alerted to the operation. At around 9.30 p.m. pumps were again increased to eight to allow a foam blanket to be placed over the leaking product. Those officers in attendance working with company staff had also estimated that some five tons of ethyl chloride had leaked and was now evolving hydrochloric acid gas. In order to place the foam blanket, the use of the water monitors was discontinued. Again this is often standard practice since the water curtain would nullify the effectiveness of the foam by making it unstable and even transferring it to another part of the site. Losing cover in this way can present problems; the gas evolution can temporarily increase placing those close in, at risk and requires a degree of technique if it is to be successful. This is therefore a tense time for those involved and one action not taken lightly even though it is routine. The incident commander has to be both alert to these possibilities and have good overall co-ordination with site managers as the change in tactics unfolds.

It is worth noting that during this period of the operation, works personnel with firefighters had entered the leak area in an attempt to both identify the source of the leak and isolate it if at all possible. In recognition of what was a developing incident, various officers of command rank had subsequently attended the incident, either at the Work's Emergency Headquarters or Forward Control, so replacing other various officers by assuming command of their part of the incident. All this activity, in command turnover, took place in less than one hour of incident duration. There is always some risk that in handover factual information is lost as task priorities impact upon individuals working at high personal stress levels.

Getting the Facts Right

The most critical aspect of this change of command was that whilst attempting to reduce the off-site impact, using water spray curtains, little had been known of the highly flammable nature of the original leaking product. The crude ethyl chloride reactor liquor that was involved was approximately 90–95% ethyl chloride with up to 2% hydrogen chloride. The main hazard was ethyl chloride although the hydrogen chloride is toxic and corrosive readily forming hydrochloric acid mist on contact with moisture. The ethyl chloride is primarily a flammable risk and, although its boiling point is close to normal ambient temperature, it is generally handled as a liquefied flammable gas when it is in its liquid form. When ethyl chloride burns, it produces toxic fumes as part of the combustion process and a pool fire would both produce high thermal radiation and toxic fumes. This information was not, however, either available from technical information on-site with the fire-fighters or offered by the technical staff available at the incident at that time.

The difficulty for the initial fire service incident commander was that he was unaware that this was the primary hazard. A very important point, when it is realised that it was the flammable gas alarms sounding at around 8.20 p.m. that alerted the works to the leak on the ethyl chloride plant. On arrival the risk information cards and available expertise from the site fire service all suggested a toxic hazard. Gas was moving off-site. A visible cloud existed. It was large and increasing. Activity in the company control room was high and, to the uninitiated, confusing. The tactic suggested of deploying a water curtain was confirmed. On the ground the physical effort was demanding in noisy and threatening conditions. Personal protective clothing gets in the way of speech, vision and hearing. The water curtain was in place and now, at the suggestion of company advisers, it was to be replaced with a stable foam blanket. All was going to plan to handle a toxic leak, not a flammable one.

It was with some surprise to those initial personnel on-site then that at just after 10 p.m., the gas cloud ignited with considerable force. The source, as the subsequent Health and Safety Executive (HSE) Inquiry revealed, was probably a release at a point between fixed pipework and a discharge port on a pump within the ethyl chloride plant. The HSE concluded that this possibly arose either because of a corroded securing flange or the failure of specially designed bellows although the HSE view was that corrosion was the most likely cause. The subsequent fire was indeed very severe. It ultimately required a major foam attack involving special procedures for the entire north west of England's foam stock supported by twelve pumping appliances and the major foam tenders on-site.

Brigade Commander's View

After my own arrival at the ESRB shortly after 9 p.m. I had commenced a review of the site and product details whilst examining the major off-site implications. The confirmed product details held at the ESRB indicated fires could be extinguished with either foam or water and that toxic fumes were evolved in ethyl chloride fires. I had been advised by both the DOSEC and a neighbouring chemical company that gas was indeed going off-site. Pressure was mounting for an urgent decision. The DOSEC wanted to know whether to commence evacuation of the public and the police were very concerned about the risk to motorists on a nearby motorway. They wanted to close this and an adjacent railway. My own view is that if gas is in the air, evacuation is not necessarily the safest or soundest option. Disrupted traffic can sometimes create serious operational difficulties in the very area you are trying to keep clear as drivers experiment with unfamiliar routes. The situation on the ground suggested our original approach of water sprays and no evacuation was a good one and I wanted to press on with the tactic. Effectively this meant re-appraising the DOSEC and police whilst pressing the Works Emergency Headquarters staff for constant up-to-date information.

Then the fire occurred, as described above. My whole plan required re-evaluation. This was a major hazard site. Ethyl chloride was not the sole or major hazard. Other more damaging products were present, not least chlorine which was stored adjacent and now at risk. The almost immediate decision was to attend the site in person. Why? Because key information was required for the next series of decisions I needed to make and I considered actual observation of the visual impact would be critical. It was as it happened that I was the most experienced officer for this type of incident who was available. In the event I curbed this very natural instinct and asked the very experienced Deputy Chief Officer (DCO) to attend in the first instance while I started to put in place, at the ESRB, a chain of command to manage a public evacuation if that became inevitable.

Time is a Pressure

This was a critical time. I believed that storage tanks containing a further thirty tons of ethyl chloride were indeed threatened. These tanks were protected by an intumescent coating, which had been designed to provide protection for around one and a half hours. By 11 p.m. I knew a domino situation could begin if the protection did not hold or the fire destroyed the structural integrity of the vertical steel plant holding the storage vessels. On such a site as this, that could lead to a highly hazardous situation.

It would mean that a critical decision had to be taken at around 11.30 p.m. to commence total evacuation of all firefighting personnel from the site. As mentioned detailed consideration also had to be given to the fact that this was a major hazard site and that within forty metres of the fire was bulk chlorine storage. I could not 'see' the visual signs of what was happening only 'hear' what was happening, through often restricted official messages.

Seeing is Believing

My eyes were those of my colleagues. I could ask questions, which I did, with increasing levels of detail. What colour are the flames? Exactly how high? Was the foam stable? Did the elevated vessels look secure? Could they see what was happening to the intumescent coating? I was reassured that the major foam attack mounted was holding even though the fire's intensity was undiminished. To do this an open line telephone link was established for private conversations, but unfortunately those on-site had to keep leaving the phone to fight the fire!

My arrival at the ESRB had coincided with a particularly difficult phase in terms of this communication between the ESRB and the DOSEC. The telephone link had been established as was normal practice but officials within the DOSEC were having difficulty in interpreting some of the more technical issues which were being conveyed to them from the emergency services at the ESRB. Fire service liaison staff had been despatched to help DOSEC staff, but it was apparent that the pressures within the DOSEC were rising as local officials and politicians sought reassurance and guidance as to the need to evacuate the nearby local population.

External Messages

I had concluded at the ESRB that evacuation was not required at this stage, having determined, given the wind direction and exposure risk, that shelter was preferable.   The time of day of the incident meant that news broadcasts, including national broadcasts, were now reaching a stage where a great deal of information was being conveyed by the media to the public. This was heightening their fears that something dreadful was about to occur.   Naturally a concerned public were then seeking reassurance since they were aware that this major hazard site had the potential to create very real danger to them.   Unfortunately the reassurances being made to the media particularly in terms of the control on-site and facts such as wind direction were not being given the same profile as the risk and potential for harm.

This had also given rise to increased concerns regarding the nearby motorway and railway links that were still operative, which again had the potential for placing many people within a close proximity to the major hazard site. As the fire brigade commander I therefore found myself in the centre of a discussion that was more to do with reassurance and communication than tactical or operational decision-making. This at a time when I needed reassurance that the tactical plan was working.

Great reliance also had to be placed upon the company official who had attended the ESRB, although he was finding that a confusing picture was emerging as to what was actually occurring on-site.   The consequence, despite further communication to the company site, was that I was finding it hard to draw a perspective on the potential for likely harm or development of the incident.

This lack of a clear outline of how the incident was developing became further confused when the information relating to the intumescent protection provided to the storage facilities arrived at the ESRB. This gave rise to a concern already touched upon that there was a real potential for further loss of containment of the stored materials if the intumescent protection failed.

This information was passed to the DOSEC to give a clear indication that whilst evacuation was not currently required there was a possibility that it might be needed within a time frame of less than one hour. Inevitably this then meant that the DOSEC had to start making preparations for evacuation and the police equally had to consider closure of the nearby motorway and railway links.

Undertaking such operations involves major disruption. Inevitably, the media became aware that the emergency services were preparing for an even worse situation. As already mentioned, the news was going out on a regular basis. Given the situation I had also taken the decision that it would be important to commence closedown operations on adjacent company sites that were also handling major hazard materials. These companies were, therefore, alerted and advised to commence a process that might involve the evacuation of large sectors of their own operations.

Taking this sort of decision cannot be undertaken lightly given both the high economic value of the materials involved and the need for the operations of closure to be taken in a very sequential way to ensure that they are done safely. It is true to say those having to shut down an operation have to trust the judgements of the fire service commander.

Finding a Solution

Whilst these external decisions were being made, further attempts on-site had been made to verify the product's properties both in terms of its exact state and quantities. This was essential information if any attack was to remain effective and equally to enable assessment of the likely impact of a catastrophic failure of containment. Attempting to assess the impact of the chosen tactics gave rise to the further concern. The foam attack, which had now commenced, was likely to be insufficient to either fully extinguish an open fire or to provide protection for other parts of the affected site which might become involved should a catastrophic failure occur. The adjacent chlorine tanks, for example, required adequate protection since their product could not be discharged safely to other parts of the site or road tankers within the existing time frame.

The decision to continue with a major foam attack also meant that major movement of bulk foam stocks had to commence from the whole of the north west of England. Whilst this was a pre-planned procedure, the logistics to achieve the desired impact effectively involved doubling the resources of appliances and personnel allocated to the incident so doubling the spans of control and command. This was to allow the transfer of stock and to contain it within the tanks of appliances so that it could be moved rapidly on-site if required.

Fresh Assessment

To help in these assessments the DCO, who had proceeded from the ESRB to the on-site works emergency headquarters and then on to the Forward Control position, was able to better inform me as to exactly how the deployment might be achieved. In addition the DCO was able to send first hand information back to the ESRB so that I was in a better-informed position regarding the whole scenario. This now included effective media management, preparations for evacuation, logistic support for a major foam attack, overall control of information regarding the site and the potential for development, and the formulation of tactical plans designed to bring the whole situation to a safe conclusion. Keeping in touch, moving the plans forward and ensuring the whole process was co-ordinated was my job.

This debate and discussion at the ESRB continued until the critical time period when it was likely that the intumescent insulation might fail. At that stage, sufficient resources had been assembled on-site to commence a major foam attack, which it was hoped would extinguish the main fire and so provide the level of protection to prevent the failure of the storage vessels. This was therefore the point on which the whole operational strategy and tactical plan was to balance.

Look for Yourself

I then made the decision that it was appropriate for me to attend the site so as to better assess the situation and gather at source any further information available to help in determining whether to request the evacuation and shut down of such a large area. In moving to the site from the ESRB, I had ensured that the necessary logistics officers were in a position and able to act quickly should there be the decision to alert the other emergency services in the DOSEC to move to the evacuation phase.

Arrival on-site confirmed that there was indeed a serious situation. The positioning of major foam monitors and cannon had commenced with the DCO now in charge of launching the attack. I made some minor adjustments to the attack plan and the foam attack was then launched.

After about fifteen to twenty minutes of this major foam attack I assessed that it had started to achieve its objective and progressively the fire was reduced both in volume and intensity. It was however not possible to extinguish the fire at this stage due in part to the damage that had been sustained to the plant and the fact that the storage facility was elevated above the damaged section. This created a siphoning effect so enabling the fire to keep going for some time. It was my estimate that the severity of risk, particularly to other exposed parts of the plant, was containable now that the major resources of foam and personnel had been assembled.

I remained on-site for some time assessing how this attack was proceeding until satisfied that it would be safe to further advise the other emergency services and DOSEC that it was unlikely that an evacuation would be needed.

It is fair to say that a degree of relief was evident amongst all those involved once it became apparent that the attack had sustained a reduction in fire severity. It was not possible at this stage to estimate how long the fire would continue to burn. The entire emergency operation had to be sustained well into the night and early morning.

So what were the Lessons for the Fire Brigade Commander?

• Information is Critical to Success.

Clearly, there was inadequate and ambiguous information about the nature of the hazards, which affected the appropriateness of the response that the fire brigade was going to make. The brigade, had it had access to its own information sources of some detail, may have been able to verify the information that was given to it. Use of the fire service's primary chemical hazard information source, the emergency action code, was not helpful, giving only the protective clothing and evacuation requirement not the primary flammable hazard. Each of the successive fire service commanders on-site had been able to speak to very well informed company officials. It was probably understandable that each should take the information that was offered at its face value, as being the best practical information upon which to base operational tactics. Even where those tactics involved placing personnel within an extremely hazardous area.

• Risk Assessment is Essential.

Again, and this incident it must be remembered precedes the current fire service methods of dynamic risk assessment (Home Office, 1999; HSE 1998, and see Arbuthnot; Crichton & Flin this volume), the risk assessment that was actually taken was more linked to trying to prevent an off-site incident than to contain the on-site effect. Cloudburst itself is a plan brought into operation by any incident, where there is primarily an off-site toxic hazard risk. Initiating the plan again had the impact of directing attention towards a toxic risk when the main hazard as we now know so well was a flammable one.

• Pre-planning is Necessary for Success.

The fire service is not necessarily in the position of holding or retrieving the sort of information that a multi-risk complex chemical site like Associated Octel requires. The procedural operational arrangements for such incidents did work and followed their prescribed form. A Forward Control, Works Emergency Headquarters, an Emergency Service

Reinforcement Base and a District Off Site Emergency Centre were all established.

- Delegating Must be Effective.

As the brigade commander I found myself in a position of having to support these four working centres whilst trying to maintain a strategic overview of the developing scenario. The risks and considerations that had to be balanced were the resources deployed on-site, the protection of on-site facilities, the threat on-site to other facilities and the threat from the on-site incident to the community off-site (which included other complex chemical and populated areas and a series of other queries relating to movement on motorways, railway links and environmental concerns arising from airborne and surface pollution).

- Communications Must be Robust.

The scenarios unravelling themselves on-site were moving extremely quickly. Within one hour the incident had escalated from a flange leak to a major spill to an ignited fire. Within a further hour it had threatened to involve significantly greater storage and indeed much wider toxic risks. The decision processes required strong communication links between all the working centres and the source of primary information about the hazard, the company.

- Balance Skills and Judgement with Needs.

Working as the Brigade Commander I sought to exercise the command process from the emergency services' reinforcement base (ESRB). Given the scenarios that were emerging, I then found it necessary to attend the site to assess for myself the likelihood of success of the major foam attack when it was launched at around 11.30 p.m. Was I right to move to this location at that time?

Hindsight suggests that it was the right decision, given the critical decision required – to evacuate or not – since this was the only location from where it was possible to make the final assessment. The management of the ESRB could be left in very capable hands so my departure from accepted practice felt safe and offered the right personal balance of responsibility and accessibility.

- Shelter is Better than Evacuation if you can Achieve it.

Whether or not to order a complete withdrawal of fire brigade and works personnel and issue the major evacuation order for a significant population that included the elderly and the young was a decision of some weight. It was likely to be challenged in court if losses, financial or human, occurred.

- Recognise the Political Requirements.

Equally political pressures would have been immense post incident, had the wrong decision been made. Part of the judgement had to be whether to rely on and so place the responsibility for the decision on more junior officers. Would this have created for them a fairly significant dilemma? After all, are brigade commanders not expected to exercise ultimate operational judgement in the fire service?

- Corporate Liabilities are Real and Have to be Considered.

There is also the dilemma that had the initial decisions of the responding commanders resulted in the death or serious injury of those fire-fighters sent forward on reconnaissance duties, there would have been a Fatal Accident Inquiry. They had acted in good faith and with the information available. Their role of leadership and delegated responsibility placed them in a no win situation, since to ignore the advice on-site and follow another series of tactics required scientific knowledge and chemical process awareness neither they nor I have. I was likewise extremely conscious that undertaking the foam attack would expose fire-fighters to high risk and I wanted to know that my decision was the correct one. So I attended the site; evidence perhaps that we all feel vulnerable. As it was, the Health and Safety Executive Inquiry concentrated upon the incident primarily because of questions raised as a direct result of public concerns.

The Health and Safety Executive did subsequently prosecute the company and on the 2nd February two years after the original incident, contraventions under the Health and Safety at Work etc Act (1974) were found proven, the company having put at risk the health and safety of employees and other people, in particular the fire-fighters involved. A fine of £150,000 was levied together with costs of £142,655.

- Professional Judgement is Critical.

There is little doubt; the fire service commander played a critical role in the decision processes of this incident. A poor tactical decision had been made initially, due to an information failure. Had the right information been available, it is possible that the fire might not have occurred, although this must remain conjecture. Once the fire had occurred, again, information became critical in forming assessments of how the situation was likely to develop. Upon those decisions were based not just issues of safety but also health and environment. Throughout the incident, fire service commanders at various levels were attempting to mitigate a leak and control the fire. Their decisions were economically important and safety critical. Ultimately, there was little off-site damage. Likewise, on-site the strategy and tactics resulted in good effective control which prevented spread with little damage occurring other than to the affected plant.

• Personal Protection is Essential.

The classic dilemma therefore, arose for the fire-fighter to stay and fight the fire in what was a hazardous situation or to move further away reducing personal risk but perhaps heightening the risk of spread. All actions, which could be taken to mitigate personal risk to fire-fighters, were taken. Clearly the wearing of personal protective equipment was mandatory, so also was the use of equipment which enabled covering jets and ultimately foam monitors to be used semi remotely thereby reducing personal risk.

• Recognise the Impact of Stress.

All of these considerations were made in stressful situations in incident centres where noise from various communications were incessant and where questions were constantly being asked which required urgent if not immediate responses.

• Know Your Operating Environment.

At the incident scene, the noise of the leak and the fire, with the roar of the pumps and the monitors in the dark of a February evening provided its own external environment of pressure and stress. The nature of major complex chemical sites is that those who are not used to that operating environment feel threatened simply because of the noises of the manufacturing processes. Any minor variation in the background noise is, therefore, carefully listened to when a major incident is in progress.

- Listen – Watch and Learn.

You find at incidents like these that you have to attempt to recognise whether this or that new noise or change in flame or depth of mist or gas is an indicator of a change in chemical process that may be presenting a higher magnitude of risk to colleagues.

- Analyse Everything Available.

Remaining calm, analysing data, deploying resources, measuring effectiveness of firefighting actions under such circumstances is the textbook response. Skill, combined hopefully with good information, and sound-working practices also helps! Incidents such as the one described indicate that, on occasions, a bit of luck can also be a major component in avoiding further injury. There is often a blind side during incident command situations and eradicating that blindness, is helped by transferred information, knowledge and experience.

- Learn Each New Lesson.

One has to ask whether, given current technologies and approaches, a better decision process can be introduced which minimises these risks and helps to avoid placing the fire service commander in such circumstances. This becomes all the more relevant as opportunities to experience large or complicated incidents reduce. My own development benefited from a number of these types of incident where I was led by people who had dealt with similar situations before. They helped ensure my own knowledge was improved whilst I gained confidence and a healthy respect for the chemical industry. I hope this note assists in that learning process.

This incident reinforces the need to:

- Check and use all sources of information. Never assume;

- Develop an objective and professionally sound operational plan. Stick to it unless the objectives change; and

- Don't be afraid to vary a procedure if that improves the likelihood of success or enables better decisions to be made.

References

HSE (1996) *The Chemical Release and Fire at the Associated Octel Company Ltd.*, Sudbury: HSE Books.
HSE (1998) *Fire Service Guidelines*. London: HMSO.
HSE (1999) *A Guide to the Control of Industrial Major Accident Hazards Regulations.* Sudbury: HSE Books.
Home Office (1999) *Fire Service Manual (vol. 2:) Fire Service Service Operations. Incident Command.* London: HMSO.

7 Military Commander – Royal Navy

Jeremy Larken
Operational Command Training Organisation

Jeremy Larken is Managing Director of OCTO Ltd, which he formed in 1991 following the Cullen Report on the Piper Alpha disaster. The company played a leading role in the ensuing emergency management developments in the oil and gas industry, with associated advisory work and company-wide command and control measures. This extended to corporate crisis management on a broader industrial and commercial base, currently involving Turnbull compliance measures for the management of corporate risk. Major clients include the North Sea oil and gas sector, UK nuclear power, chemical, pharmaceutical, electricity generation and telecommunications industries, and a Ministry of Defence agency. Larken has tutored, counselled and reported upon over 1000 senior managers, including a number at board level, many with their emergency response teams.

During his previous 33-year naval career, he held six major sea-going commands, highlights of which are summarised in the text below. He was awarded the Distinguished Service Order (DSO) for his contribution to the Falklands Campaign.

www.Octo.uk.com

Introduction

The canvas I have selected is the Falklands Campaign in 1982, events encompassing some 100 days. My post was Captain of the 12,000 tonne assault ship *HMS Fearless*. In addition to command of the ship, I was Chief of Staff to the Commodore Amphibious Warfare (COMAW), the maritime commander for the military landings at San Carlos – the pivotal step in the re-possession of the Falkland Islands. As such, I was involved intimately with the planning of the amphibious tasks and the naval inshore operations around the islands, and the Commodore assigned me extensive elements of delegated tactical command authority within the amphibious force. Drawing upon the sum of my previous command experience, I could scarcely have been better prepared for the challenges; nor could I have gone to this peculiar war either with better people or in a better ship. Then 43, my first major command, a fine diesel-attack submarine with a crew of 60, had come to me at the age of 30. I had spent 12½ of the next 18 years in six command appointments of increasing responsibility and complexity. Three were in the submarine flotilla during the height of cold-war operations, and I was fortunate to keep an equal footing in the surface fleet. I stepped finally into the shoes of COMAW, commanding the UK's amphibious forces (and also, within NATO, those of the Netherlands), from 1985 to 87. On concluding my service with the Royal Navy in 1990 as a Rear Admiral with a central role in the Ministry of Defence crisis management organisation, I founded OCTO Ltd. (see above and Larken, 1992, 1995a,b).

Scope

The Falklands crisis broke as an almost complete surprise for those who were to fight. *Fearless* abruptly found herself a major cog in a large and intricate yet flexible military machine that quickly became a crucial instrument of UK national will and government policy. Mine was not the hand that directed the strategic destiny of *Fearless*, much less the Amphibious Force. But I had everything to do with preparing the old ship for her starring role, a process that you will find stretched back a year. During the campaign itself, you have heard already that many practical decisions concerning the Amphibious Force as well as *Fearless* fell to my direction and execution. I was moreover responsible for all souls on board my ship, some 1400 for most of the period. Since the quality of *Fearless*'s preparedness proved crucial, I shall describe how this was achieved, before covering illustrative elements of the campaign itself. In the context of the narrative, there is a section on practical decision making, and some reflections on the nature and exercise of command. My chapter concludes with a summary of principal lessons.

HMS Fearless

Fearless was an amphibious assault ship. Completed in 1963, remarkably she is still (2001) active as by far the oldest operational ship in the Navy, having been seen as quite an old lady aged nineteen at the time of the Falklands Campaign. A key feature is a stern dock, the size of a substantial swimming pool. Into this fitted neatly four large landing craft. At the forward end of the dock is a metal 'beach' which accesses various parking decks, similar to a roll-on/roll-off (ro-ro) ferry. These accommodate vehicles up to battle-tank weight. Four smaller landing craft operate from davits. Much of the stern area above the dock is covered by a flight deck, which, at a pinch, can operate without hangerage four large Sea King helicopters – and on one notable occasion in San Carlos a fully armed Sea Harrier jump jet.

Figure 7.1 *HMS Fearless* Docked Down in San Carlos Water in her Multi-role Battle Array

One of the large landing craft can be seen in the stern dock entrance, and two small ones are suspended from davits on the port side at main-deck level. A Sea King helicopter is operating over the bows. This was the single occasion when a Sea Harrier landed on the flight deck to refuel. The two black globes on the mainmast house satellite communications. Just to the left of the offset forward funnel (which is in line with the foremast) can be seen the weapons-direction platform from which the author of this chapter conducted the battle. (Photo: Captain J. P. Morton CBE., M.N.)

The operations areas further forward in the ship are exceptionally extensive. They incorporate three major command and control facilities. In addition to those required to fight the ship, there are headquarters for a maritime amphibious commander and for a land-force commander. These two guest commanders and their staffs act in an orchestrated relationship that does not concern us here beyond flavours of my personal involvement. The facilities themselves were obsolete in 1982, with the exception of the long-haul communications and – surprisingly – a closed circuit television (CCTV) system.

The permanent ship's company comprises some 600 men, of whom around 90 man a Royal Marines Assault Squadron that operates the landing craft. From the Falklands, six, but only six, to our grief failed to return. We could then accommodate, in reasonable comfort, a further 400 men as an embarked force that could land from *Fearless* direct. Somehow we expanded our hospitality to some 1,400, many short-term lodgers, throughout the campaign. For 36 hours before the San Carlos landings, we peaked above 1,700.

The Bolt from the Blue

The day was Thursday 1 April 1982. Having returned from her first operational deployment, *Fearless* was one week into a three-week shutdown period for maintenance. Much equipment had been opened up, nothing reassembled. Dark news of the invasion of the Falkland Island by Argentina distressed an evening meeting with some of my managers. Although this was just the kind of situation the Amphibious Force and *Fearless* had been designed to redress, we did not imagine that dear old UK Ltd would have the 'bottle' for such a seemingly quixotic enterprise.

We were wrong. Next morning I was awakened early by a call from my second-in-command, Commander John Kelly. *Fearless* had been placed on priority for war stores second only to the submarine strategic deterrent; perhaps I should hasten onboard. That Saturday between briefings I heard by radio Michael Foot, left-wing Leader of the Opposition, help Margaret Thatcher call the nation to arms in the House of Commons. Alan Clarke (1994) described as 'the most electric moment that I have ever experienced in that place, or it for many years I suspect, perhaps since 8 May 1940' (p. 337). On Tuesday we sailed from

Portsmouth for the South Atlantic with the heaviest and most varied load the old ship had ever embarked.   The Amphibious Commanders, Commodore Michael Clapp and Brigadier Julian Thompson, joined by helicopter in the Channel. The Amphibious Force gathered around us and we set course for Ascension Island, 4,000 nautical miles away near the equator and some half way to the Falklands. Many more ships, men and equipment were to follow in our wake.

The Foundations

Looking back to the beginning, a major lesson re-learnt is the enormous importance of operational preparedness for an emergency. I suggest that this involves four vital pillars:

- a sound strategy prosecuted consistently;
- a robust business plan;
- facilities fit for purpose, resourced sufficiently; and
- a motivated, trained and integrated management and workforce, united in their belief in what they are being invited to do.

A commander may lack these important ingredients.   This will be a grievous handicap, and there will be corresponding limits to what can be accomplished. Even Napoleon ceased to be effective when finally shorn of assets and communications on the island of St Helena. Consider the much more recent collapse of US moral purpose in Vietnam.   Superpower became impotent. Huge segments of British industry have disappeared following the erosion of these pillars, to be replaced by more-relevant entrepreneurial endeavours. Any great undertaking that fails to prosper, or which crumbles under pressure, will probably lack at least one pillar.

A Year of Preparations for the Unexpected

Setting the Strategy and Driving the Agenda

I took command of *Fearless* in May 1981 on relinquishing command of a squadron of nuclear attack submarines, a year to the month before the San

Carlos landings. She was a sorry sight, hugely high and dry in a floating dock in which she was undergoing a major overhaul. The lead contractor was a struggling Tyneside ship repair yard, caught in a fixed-price Ministry of Defence contract, the successful completion of which might mean more work and survival.

The Royal Navy possessed two assault ships. Customarily, one had been active whilst the other languished in reserve status – from which she would in due course undergo an overhaul before once again taking the lead role. Thus *Fearless* was scheduled to take over from her sister ship *Intrepid* during the 1981/82 Christmas/New Year break. This would entail the transfer of substantial equipment and a number of specialist personnel. Taking stock with my new team on the Tyne, the chances of *Fearless* meeting this schedule looked frail. Such doubts had percolated to Fleet headquarters. There was ominous talk of deferring the overhaul's completion and extending *Intrepid*'s tenure as the active assault ship.

I needed to make early judgements and decisions that were realistic and responsible with an incomplete information base. I held two trump cards. First, to extend *Intrepid* would be highly unwelcome in manpower management terms – her people were earmarked already for new posts. Second, the shipyard would try desperately to complete *Fearless* on time, even if engineering standards were threatened – a danger we must circumvent. Having done my utmost to galvanise and focus the admirably pragmatic Tynesiders, I needed to convince my own masters it was safe to plan *Intrepid*'s timely release to reserve status. I knew that as time passed the *Intrepid* extension would fade as a practical option, and I would then more readily be assigned 'tiger-team' resources to deal with the serious critical-path deficiencies that I had no doubt *Fearless* would experience.

The strategy in place and nurtured from week to week, I turned my attention to the timely completion of the overhaul and the development of the ship's company. The latter was a matter of hearts and minds, and training. In John Kelly I had a splendid second-in-command and deputy, following a heroic career as a commando helicopter pilot - firm, intelligent, mature, gently gregarious and impeccably loyal. Delegating to him and other departmental heads extensively, I was free to deploy my personal resources where most needed.

Establishing the Credentials Underpinning Leadership

The role of an amphibious ship demands close Royal Marine associations; for instance one-sixth of the ship's company comprised Royal Marines. I knew little of this illustrious corps, but reputedly sailors and marines were not natural kin as colleagues afloat; no more were surface sailors ('fish-heads') and commando naval aviators ('junglies') intuitive soul mates.

This submariner was determined that *Fearless* should not become a floating colony of professional ghettos. To promote entente, I set up two weeks' training at the Royal Marines Training Centre at Lympstone, Devon for the sailors in *Fearless*'s deployable platoon (diminutive great grandson of the Naval Brigades of WW I) in the rudiments of field-craft and infantry weapons, to be instructed by marines of our land-craft assault squadron, and went along myself. Having taken precautions to get fit, I was better placed than some of my toiling, sweating sailors in route marches, assault courses and other delights, establishing credentials both with them and my marines, and getting to know all concerned.

I also undertook a week of pilot training with the naval commando helicopter squadrons at Yeovilton. Awarded a huge pair of cardboard wings for successfully holding a helicopter in the hover for six seconds, my solid professional attainments were unspectacular. But here again I built partnerships that were to stand me in the utmost stead.

There were many other people to get to know at all levels. Whenever back with *Fearless* at South Shields, I explored the ship ceaselessly, often at unconventional times of day and night. In the mutual interests of timely completion, the shipyard agreed that sailors could undertake some hands-on work without union censor. Meeting my workforce of sailors engaged in basic engineering, ship husbandry and painting tasks, often boring, provided matchless opportunities to learn, to inform and to build mutual confidence and common purpose.

Achieving Intimate Personal Knowledge of Fearless, as a Baseline to Establish my Command Credentials and 'pour encourager les autres'

In my first command, the diesel submarine, my over-enthusiasm once led to what I judged to be a potentially mutinous situation. I averted the possibility with direct briefing of my people, and learnt richly from the experience. One source of preparation for this volatile incident was the observation of a senior colleague, some years previously, handling stressed sailors with what I reckoned to be a peremptory lack of sensitivity. This truly distinguished officer, who rose to very high rank, never developed a natural touch with his people. In walk-about mode, he was uncomfortable and rather bored, whereas I found the benefits enormous from early days in the Navy.

For me, skills in walkabout management developed partly from instinct and partly from the submarine culture. Whilst not lacking elsewhere in the Navy, submariners tended to have walkabout management forced upon them, even very large submarines imposing sardine working and living conditions by any normal standards. Those of us fortunate enough to spend much time also in surface ships, brought our relaxed peripatetic habits with us.

In my experience, this is an invaluable approach, and I consolidated its practice in successive sea-going appointments. Early on, no-one expects the boss to know everything. The licence to ask questions whilst exploring every extremity of the ship is thus unlimited; and people love to talk about their jobs and themselves. The habit of asking questions can then simply continue, with due sensitivity of course, even if people might have suspected my purpose had gradually become reversed! It also stimulates the rest of management. Provided it is clear that the process is constructive and not an invasion of departmental turf, most managers will be accept the challenge cheerfully and simply take care to keep ahead of the boss.

There was a question of universal basic ship knowledge. Imposing my submarine instincts, and reinforced by John Kelly's as an aviator, I insisted – against considerable traditional opposition – that everyone must achieve an accredited standard of whole-ship (as opposed to simply departmental) knowledge. I rounded my own education with a few days at sea in *Intrepid* under the tutelage of Captain Peter Dingemans.

Thus were my personal credentials to command established, and the motivation, competence and integration of the *Fearless* workforce advanced by me and my colleagues, even in the unpromising environment of South Shields with its fantastic density of pubs per square mile and other distractions. These qualities proved vital in extricating *Fearless* from the valiantly struggling shipyard just about on time, ready to navigate the shifting sandbanks of trials, some failed trials and ever-tauter get-well deadlines. We achieved clearance to move forward to safety and operational 'work-up' with the narrowest of scheduled margins, and a number of tolerable compromises - by which time the ready extension of *Intrepid* was impracticable.

The Decision Making Process

I have been asked to provide some observations on decision making. In doing this, I shall try to relate my own experience to some of the decision-making approaches advanced in this book (see Crichton & Flin), allied to some distinguished historical provenance that seems relevant. Throughout the period described so far, I orchestrated an elaborate decision-making process, consistent with the strategy we had established whilst sensitive to ever changing circumstances.

Analytic Decision Making

We are all familiar with this systematic approach, generally undertaken without great time pressure. It has a high formal standing in British military practice, in the shape of a process known as 'Appreciating the situation'. General Montgomery is reported to have ordered several independent appreciations in preparing for the battle of El Alamein before settling his battle plans, although I cannot find a reference for this. Many appreciations were written to support elements of the Falklands Campaign, and in some I had a hand (see Clapp & Southby-Tailyour, 1993). As practised by UK forces, I rate 'Appreciating the situation' as an immensely powerful aid to logical thought under pressure, and I recommend it strongly to our business clients. There is an obverse to the coin: 'Situating the appreciation'. This quip recognises possible tendencies to decide on the solution and then justify it by any means that time and energy allows, perhaps after the event! This could, I believe, spill over into misuse of the naturalistic approach, even for the best of reasons.

Naturalistic Decision Making

I am indebted to Dr Gary Klein (1998) and Professor Rhona Flin (1996) for my introduction to the modelling of naturalistic decision making in relation to its better-understood analytic complement. For the naturalistic zone, Gary Klein coined the term Recognition-Primed Decision making, or RPD (Klein, 1997); Admiral Sandy Woodward calls it 'picture matching' – like summoning a picture from your memory and calibrating and adapting it against what you see. With colleagues in OCTO, I have thought much about the applications of analytic and naturalistic decision making techniques, and the spectrum of approaches that links them. We have tested these methods in the light both of our past experience and the various situations posed and demonstrated by our clients at many levels.

The proposition of RPD is that it is indeed recognition-primed. This creates the self-evident dilemma in emergency and crisis situations, whereby the decision maker is liable to be operating in a fast-moving and often malign environment which is beyond his or her previous experience, and hence beyond assured 'recognition'. Under such pressure 'the first solution that will suffice' (judged a satisfactory RPD outcome) may all too readily become the (untested) 'first solution that comes to mind'. The dangers are obvious, and the disaster in which the United States Navy's cruiser Vincennes erroneously destroyed an innocent Iranian airliner is a dreadful example (see Salas et al. this volume).

Yet the fact must be faced that the emergency or crisis decision maker needs to be able to make vital decisions beyond previous experience, against an assessed (or perhaps just sensed) time-base which does not provide opportunity for deliberate application of the analytic method. There is no escape here. No panacea of a sure formula is on offer.

There are however some guidelines, and OCTO teaches these, with a variety of models, some developed by us and others by Professor John Strutt (1997) and colleagues of Cranfield University. First, starting with a sound analytic structure, aspects of RPD can be plugged in and some analytic corners cut with a degree of confidence. Second, much can be done by maintaining the strategy and plans derived analytically, and thinking through action contingencies ('what if-ing') to be applied on an RPD basis. The first long-proven Principle of War is 'The selection and maintenance of the aim' (Royal Navy, 1995). There is something also to learn from the various aphorisms about luck, of which I find 'luck favours the prepared mind' the most telling.

For the Royal Navy there is nothing new in this. These are indeed the famous Nelsonic principles, conveyed to individual commanders as colleagues (the 'band of brothers'), and cascaded down. The inheritance is the essential understanding that 'the plan does not survive the first shot'. Commanders, at all levels, should sustain the strategy and overall plan whilst using their initiative to carry the momentum of battle forward. They watch colleagues so far as they can through 'the fog of war', and act as they judge best largely without further orders or direction – maybe fragments of top-down steerage if they are lucky. This is nothing more nor less than RPD, at its best co-ordinated intuitively between colleague commanders who know the master plan and each other well, augmented when necessary by what I understand to be creative decision making (a relative to 'lateral thinking'?) 'on the hoof'. Wellington practised it, literally on the hoof, in his equally elevated way during the Peninsular War in Portugal and Spain (1808–14). The concept was adopted, directly or indirectly, and formalised by the German General Staff under Bismark, fathering the term Auftragstaktik, and used to spectacular effect by the German army in both World Wars. It had meanwhile largely been lost to sight by the Royal Navy during the Pax Britannica period (1815–c1900), leading to disappointment at the Battle of Jutland before largely being recovered in time for WW II – a process described brilliantly by Andrew Gordon (1996) in his recent book The Rules of the Game – Jutland and British Naval Command. Happily the free-ranging ethos was alive and well in the Royal Navy and the Royal Marines during the Falkland Campaign, and used to excellent effect.

As this story unfolds, you will see that there were firm strategies in place at various levels, including my own in *Fearless*, before and during the campaign, and of course in the Amphibious Force. We had the relative luxury of good time to think our way into the various unprecedented situations that were to confront us, even if things then often happened very quickly indeed. We were extremely well trained and, for nearly all the time, extremely well informed. We won the crucial information battle against the Argentines decisively, and at a very early stage. This combination enabled us to apply a high degree of RPD on many occasions, and with success. One very simple illustrative example: you will discover that I had an expectation that gun's crews might be distracted by wounds to comrades; this enabled me to act briskly when this event, beyond my experience, duly occurred.

Creative Decision Making

I am indebted also to Rhona Flin for introducing me to Dr Judith Orasanu's concept of creative decision-making (Orasanu 1997; Orasanu & Fisher, 1997). If this means what I think it means, creative decision making had a major innings on the British side of the Falklands Campaign, augmenting both the analytic and recognition-primed elements.

Creative decision making would appear to be an approach needed when the exponent finds him/herself plus team completely outside the envelope of previous experience or anticipated out-turns. It is the operational journey without maps. Let me offer some examples, mainly historical, which may interest readers:

- The campaigns of Alexander the Great – surely a sustained demonstration of creative decision-making (Lane Fox, 1973).
- The containment strategy of France under Louis XIV masterminded by John Churchill, Duke of Marlborough (Churchill, 1933).
- The maritime tactics and leadership approach devised and used with overwhelming success by Admiral Nelson, placed elegantly in a modern business context by Stephen Howarth in the Financial Times Inside Track column on the 200th anniversary (1998) of the Battle of the Nile. This approach, mentioned above in the context of RPD, was a truly original product of Nelson's, and the mark of his greatness.
- The command and control system developed by General Wellesley (later Duke of Wellington) in India and during the Peninsular War (Longford 1964).
- The degree to which there was linkage between the seminal approaches of Nelson at sea and, on land, Wolfe, Napoleon, Wellington and one or two other great generals of that period could itself be an important study. This was a period during which the concept of 'total war' moved forward, spawning a deluge of original thinking and practice.
- The 'Auftragstaktik' doctrine adopted by the German General Staff and used so successfully for a century (e.g. Hastings, 1983) does appear to be a formal and duly Germanic evolution of these principles to which Clausewitz (1832) was certainly indebted.
- During WW II, British examples amongst many are: the circumspect use of the intelligence from Bletchley Park, overseen directly by Winston Churchill and now widely on the public record, and the less well-known development of Operation Analysis by Professor Patrick Blackett and colleagues (Terraine, 1989).
- For an American WW II example, consider General Douglas MacArthur's amphibious island-hopping campaign in the Pacific and, in a completely different field, his contribution to the political reconstruction of Japan following the war (Manchester, 1978).
- What about examples from the Falklands Campaign?

I would offer:

- Admiral Woodward's handling of the small Carrier Group, keeping it to the east of the Falklands, able to hold the ring whilst more-or-less beyond Argentine reach.
- His use of the Sea Harriers. Armed with an excellent US missile (the AIM 9L Sidewinder), the very rapid regeneration rates (i.e., turnaround on deck and launch for a new sortie) enabled this small number of aircraft to hold and then completely out-class its much more numerous Argentine opponents (see Woodward, 1992).
- The decision by Brigadier Julian Thompson to send the marines and paras on foot across country following the loss of the container ship *Atlantic Conveyor* with nearly all the reinforcement helicopters (Thompson, 1992).
- The battle for Mount Harriet by Major Nick Vaux and 42 Commando, achieved by inspired tactics with remarkably few casualties (Vaux, 1986).
- I would add the use of amphibious shipping on flanking movements, but this harks back to General MacArthur – albeit an art neglected for many intervening years (Clapp & Southby-Tailyour, 1993).

These examples come from the analytic end of the scale. During the Falklands Campaign there were many examples at the operational RPD end. A well-known instance was the intervention of Lieutenant Colonel H Jones when the attack by 2 Para on Goose Green got bogged down, for which he was awarded a posthumous VC. Lieutenant Colonel Michael Rose (now General Sir Michael Rose and much decorated) was a tireless source of creative decision making (see also his book about his contribution to peacekeeping in Bosnia – Rose, 1998). This list constitutes the exploits of exceptional commanders, and it seems to me they are striking examples of creative decision making. Readers will think of others.

All this has every relevance both to business in general at the crisis management level, and to major-hazard plant emergencies in industry. You do not need to look far in commerce to find creative (for which I suspect one means 'entrepreneurial') decision-making. Jack Welch of General Electric and Bill Gates of Microsoft are obvious exponents.

Procedural 'Decision Making'

Procedures are very useful things, extensively used by the Armed Forces as well as industry. Evolution of procedures is also an invaluable way of exploring the many vital second-order effects of the issues in question and pre-canning predictable decisions. With my colleagues, however, I have doubts on the place of procedures in the spectrum of true crisis and emergency decision making. In my experience, serious decision-making starts when procedures run out; and they will, in my experience, run out very soon during an incident with 'fire in its belly'. In short, often they do not long 'survive the first shot'.

Command in Preparation for War – as it Turned Out

Working up to Full Operational Effectiveness

Back to our story and the autumn of 1981. Decision making within the framework of sustained strategy attended the ship's work-up period at and around Portland in the English Channel. We were becoming an operational team. Hard-driven, even if there was much to entertain as well as to test and exasperate, *Fearless* developed a tempered operational proficiency, with resilience to match. During the early days, under the tireless eyes of regulatory inquisitors, ancient Bofors anti-aircraft guns on the bridge wing platforms would respond to the early morning mock attacks by fast jets with sporadic blank fire before jamming. Bang, Bang........ Bang, Bang........Silence...... Woosh (went the jet close overhead)! (Six months later in San Carlos they would fire on and on with relentless efficiency.) Missiles hung-up on launchers. Inexperienced manoeuvring caused dents to landing craft. Internal damage control exercises were a spaghetti of hoses, baulks of timber, bruised elbows, lurid make-up wounds and colourful language. More and more decisions could be left to my departmental heads, with onward downward delegation. More and more was I free to pursue high-level issues. On completion of a compressed programme, we were declared operational, with some formal credit and warm private endorsement. *Fearless* had achieved her first objectives. *Intrepid* adjusted to standby status, from which she was to be re-activated abruptly three months later for the Falklands Campaign with much of her old ship's company.

New Year 1982 saw *Fearless* deploy with a small squadron to the West Indies by the northern great-circle route. This was my decision and a poor one, based on inadequate research. We encountered hurricane conditions mid-Atlantic, not unusual for the time of year, and damaged two helicopters lashed to the flight deck, one terminally and thus not available for the Falklands aircraft inventory.

After some colourful tribulations, the deployment took *Fearless* to the Netherlands Island territory of Curacao. The port incorporated an army encampment, Camp Allegro, converted to a recreation centre which proved enormously popular with my sailors and marines, the true nature of which became clear to me subsequently. The padre conceded wanly that it had taken the Camp Allegro to really put the ship's company into foremost fettle! I did not complain. We embarked for a short visit to the Commander-in-Chief, Admiral Sir John Fieldhouse. He enjoyed a traditional ship's company concert held on the vehicle (ro-ro) decks, in which five hundred voices nearly lifted the flight deck heavenwards with 'Hearts of Oak' and 'Rule Britannia'. The Admiral turned to me and said, without conceivable premonition, 'What if we were to go to war tomorrow?'

We re-crossed the Atlantic, embarked a major contingent of Royal Marines at Plymouth together with the amphibious maritime and land-force commanders, Michael Clapp and Julian Thompson, and proceeded to north Norway around the Lafoten Islands for major exercises. With the Commodore's encouragement, again without premonition, I experimented in manoeuvring *Fearless* to use huge fjordal land features as protection against line-of-sight attack by fast-jet aircraft. Driven simply by sound strategy, a succession of such decisions, both planned and opportune, advanced the ship's operational readiness and our ability to apply her capabilities in a war theatre.

Meanwhile we were learning to integrate with and support our amphibious staff guests and colleagues. My walkabout habits of leadership and second-level information management continued; a cup of coffee for the Captain in a machinery control room at 0430 continued to reward all parties.

Reflections on Command: Solitude and Conference in Command; Sources of Clear Thinking and Leadership; also the Dangers of Hubris and the Corruption of Power

Command at sea is by its nature a lonely business. It places heavy leadership demands and, if discharged effectively, has a key positive influence on the ship and her company as a whole. In principle, this reads across to the direction of any comparable unit – industrial or military. A warship however happens to be unusually self-contained, and warship command provides thereby an exceptionally pure methodology template. It provides also a very personal one. Multiple naval or military units, just like complex businesses or joint ventures, are more complex. They constitute coalitions. There will be more about these below.

In time-honoured naval tradition, a warship Captain lives and mainly works in isolated quarters. Socially, he only enters other living areas (each known in the Navy as a 'Mess') by invitation and then infrequently. This may surprise current management gurus, but it does stand out as one great strength of command at sea, proven over many generations – and, when misapplied, of eccentricity or worse! Solitude allows space for reflection and detachment, both essential ingredients in maintaining the big picture, and for ensuring a sound balance between the human and material elements of the business. At sea there is an obvious need for a particularly high level of alertness, unchanged in essentials since Nelson's day and described so admirably by C S Forester and Patrick O'Brien, and in current practical terms made the more acute by the short-fuse capabilities of modern weapon systems and their antidotes. For generations the buck has stopped at the RN Captain's door, in a manner Nigel Turnbull (1999) and his colleagues would entirely approve. Solitude links back also to my observations on the decision-making spectrum between the analytic and the naturalistic (RPD). It gives essential time and space for what-iffing. It provides a fertile seedbed from which RPD can flourish when things happen quickly, founded upon thorough prior analysis constantly reviewed. To do this well is one of the ultimate challenges to the captain of a ship, and I do not see anything different in essentials between this and the highest challenges to a Captain of Industry. As in any business, the balance between professional detachment and a vital intimacy with one's Heads of Department is a nice judgement. By conferring and informing warmly and with close mutual trust, yet standing back with a high degree of independence, and without any suggestion of fear or favour, the command can give considered direction and timely inspiration which may make the difference between success or failure in the fog of war, or in the fog of commercial opportunity or crisis alike. It seems to me that the pursuit and application of this complex balance, almost an art form, is under-valued by some modern business managers, and I fear this to be a serious deprivation.

There is an important complement to the solitary element of the command profile in the Royal Navy, and that is the custom whereby ship's captains seek out each other's company to discuss command practice at every opportunity – an approach again instituted during the Nelsonic era, possibly by Admiral John Jervis, later Earl St Vincent, and First Lord of the Admiralty and the guiding naval strategist for key periods of the wars against Napoleon. (Advising on the prospects of the French armies invading England in 1804 during the defensive campaign which culminated in the Battle of Trafalgar, he remarked dryly 'I do not say they cannot come, only that they cannot come by sea'). John Jervis saw Nelson as a prodigy, and rejoiced in his famous iconoclastic initiative at the Battle of Cape St Vincent where Nelson captured two Spanish battleships with highly unconventional tactics – a shining example of creative decision making, incidentally. Whenever two British warships are in company, the junior captain will seek to visit (or 'call upon') the senior, a custom linked traditionally with entertainment to lunch or dinner (see again Forester and O'Brien for the Nelsonic era – the custom is as lively today). This is not a pompous practice in the least; it is a matter of two commanders conferring. Thus will the captain of an aircraft carrier entertain as a colleague the immensely junior captain of a minesweeper. The latter will be received by the aircraft carrier's second-in-command (also greatly his senior) and be escorted by him to the captain's cabin.

One cardinal and appealing feature of leadership, anywhere in my view, is to be both able and prepared to do competently anything one asks of one's workforce, especially if unpleasant or hazardous or, less glamorously, plain boring. There are obvious practical limits in applying this principle: the correct balance between sharing and example on one hand and plain effectiveness on the other always requires judgement. In Egypt in WW II, General Auchinleck insisted that staff officers share fully the privations of the army's desert camp. His successor General Montgomery, taking command before the Battle of El Alamein, moved them to more comfortable quarters to work more efficiently (De Guingand, 1947; Montgomery, 1958). Yet sound reasons such as this can readily degenerate into *droits des seigneurs*, a perennial worry in military organisations – and many others.

To illustrate both the sharing and hubris dilemmas, here is a personal *Fearless* cautionary tale. John Kelly expressed his anxiety that I might overdo my enthusiasm for the aviation community and their helicopters. It was appropriate to have drawn full flying gear. Once accustomed to its use, no doubt I began to look the part. One day a new mark of helicopter landed on our flight deck in the hands of an eminent test pilot. Would I care for a ride? But of course! Duly clad, I hoisted myself into the left-hand seat like a pro and strapped in. Would I like to take the aircraft off? To me this meant following the pilot through on the controls. It did not occur to me that this paragon might think I could fly, nor to him that I was not a qualified pilot – a prime example of reciprocal flawed recognition-primed decisions, in the guise of mutual assent. Moreover he had inadvertently switched out the automatic stabilisation. During the seconds that followed we, the flight-deck crew and possibly many others were lucky to survive the near disaster of an otherwise exceedingly comic take off. I had been disgracefully irresponsible. I had posed, if artlessly, as a pilot. It is often important for the leader not to wear his heart on his sleeve. He must however take care never to act what he is not. 'Be yourself' is the most valuable command advice I ever received (Royal Navy, 1965).

It is of subsidiary interest that I was unstressed by this incident in which I nearly killed myself and many others. It is a quality of preparedness that one can accept close calls with equanimity – for a period at least.

The Falklands Campaign

All Records Broken, and a Sombre Prospect

The Falklands Campaign had its origins in a dispute between Britain and Argentina over the sovereignty of the islands, to which the Argentines refer as Los Malvinas, the complexities of which are irreconcilable. In defence of the British case, the inhabitants of the Islands in 1982 were almost universally of British stock, and their wishes as to their governance were unambiguous. Amongst civilised nations, this is the normal criterion for decisions on such matters. In defence of the Argentine position, the British Foreign and Commonwealth Office, contrary to the view of Parliament whenever it was periodically consulted, regarded the Falklands as an administrative liability and had for some years been warm to some form of condominium with Argentina – signalling to the Argentines accordingly. Thus when it suited the Argentine Junta to divert attention from an unsavoury domestic political situation with a popular military campaign to occupy Los Malvinas, it did not occur to them that Britain would do more than create a ritual international rumpus and let the islands go. This is of course not, once outraged, the British way, and the peremptory and rapacious invasion of British territory, even at almost the other end of the earth, caused outrage that was virtually universal. Firm leadership by Mrs Thatcher, armed with resolute (and to a distinct extent RPD) assurances by the First Sea Lord Admiral Sir Henry Leach that a maritime task force could be assembled to repossess the Falklands, prevailed over many faint-hearted counsels (Leach, 1993; Hastings, 1983). The colours of a united kingdom were nailed to the mast.

Fearless sailed from Portsmouth on the Tuesday 6 April with an unprecedented operational load. My people and numerous helpers, including normally rule-bound dockyard workers and the teenage sons of my Deputy Weapons Officer, moved mountains to reassemble and store the ship. Four large helicopters, artillery, ammunition, light tanks, and many lesser vehicles packed into every conceivable corner. False decks were created from a cornucopia of stores. Topping up to the 1,400 people, we broke all previous records. From the round tower at the narrow harbour entrance, crowds, including *Fearless* families, cheered and wept us on our way. We anchored immediately at Spithead and docked-down to embark our four large landing craft. Local television beamed visions of our distressed womenfolk clutching wide-eyed children. It was a sombre moment, presaging deep leadership needs even for a ship fundamentally so ready.

Establishment of Command Patterns – Less than Ideal

It is an ancient custom that the Captain of a command (or Flag) ship joins the embarked commander (Commodore or Admiral) for meals – he is in any case Chief of Staff. This I had learnt first-hand from the best of teachers, the then Rear Admiral John Fieldhouse, when I had had a windfall youthful innings as Captain of the magnificent guided missile destroyer HMS Glamorgan, then his flagship. But now, in Michael Clapp and Julian Thompson, there were two guest commanders. They would have much to talk about. I was shy of intruding gratuitously and a cautious enquiry indicated they would prefer to remain a deux. Hindsight told me I should have been more insistent. Now an adviser to company boards on crisis management, I have discovered that three or four makes a much sounder strategic debating group than two. This was, I now see, a weak decision, in consequence of which I failed to support my leaders to the limits of my knowledge and experience. Here is an example, interleaved with the enormous importance of getting the command organisation straight.

The senior commander afloat was Rear Admiral Sandy Woodward. The situation was however complicated by a command structure under Admiral Fieldhouse which appointed Sandy Woodward, Michael Clapp and Julian Thomson as co-equal Task Group Commanders, each Task Group having of course a particular function. Sandy Woodward had degrees of primus inter pares status, which varied and were by no means always clear to those concerned. Notwithstanding Admiral Fieldhouse's superb personal leadership of the campaign from his shore headquarters at Northwood, Middlesex, the consequent misunderstandings were serious. To ensure an absolutely unambiguous command and control structure is a lesson of war clear to Alexander the Great, Marlborough, Napoleon, Nelson, Wellington, and the great German Army commanders alike. Such difficulties apart, it was a shining feature of the Falklands Campaign that overall we proved so ready and so flexible, and that really very few ancient lessons were forgotten.

Having served Sandy Woodward four times, I knew him well with awesome admiration – and deep affection – as an intellectual, keenly astute and a masterful submariner by background. The Rt. Hon Margaret Thatcher in 1992 wrote: 'There were those who considered him the cleverest man in the Navy. .... He was precisely the right man to fight the world's first computer war'. (in Woodward, 1992, pxiii).

His agile brain generated bright ideas at an alarming rate. At inception these were generally not ordered, and his hapless staff could readily be run ragged without stern prioritising by a firm deputy – which it had once been my privilege to be. Michael Clapp knew Sandy Woodward only slightly and Julian Thompson not at all.

A meeting of these seagoing principals would almost certainly occur near Ascension Island where the small Carrier Group had assembled before us. Sure enough, as soon as *Fearless* was within range, a helicopter bearing the Admiral approached. The pressures from the Commander-in-Chief's headquarters at Northwood were already acute and there was a residual air of stress aggravated already by the ambiguities in the command structure.

Some days before, I had decided to attempt to prepare my leaders for the type of intercourse such a meeting was liable to bring. I was anxious above all that they should recognise the inevitable barrage of Woodward ideas, some projected simply as a cheerful agenda for lateral-thinking debate. They should stand their ground, especially on amphibious principles – on which the Admiral would wish to be educated.

Everything turned out as predicted, but my attempts to fix the chemistry proved in vain (see Clapp & Southby Tailyour, 1993, Thompson, 1992; Woodward, 1992). My two champions were defensive, and dismayed and offended by Sandy Woodward's breezy and provocatively creative demeanour. This set a pattern of mistrust, almost adversarial, between the offshore Carrier Group and the inshore Amphibious Group – the two principal headquarters groups afloat. They rubbed along together. But it was a sadly raw relationship, fraught with misunderstanding and perceived offence. I did go on trying to moderate these stresses and strains, with some modest success. Max Hastings and Ewen Southby-Tailyour, in books about the campaign, were amongst those who had kindly things to say in this respect. Max Hastings, in my view the most outstanding of a strong media corps in the South Atlantic, wrote with Simon Jenkins a book (Hastings and Jenkins, 1983) which remains one of the most powerful accounts of the campaign amongst a great deal of more recent literature. It was there with gratitude I read:

> Larken ...... was to play an important role in keeping 'combined operations' combined. His ..... great skill in sympathising with other men's points of view contributed enormously to the eventual success of the amphibious operations (p. 121).

As a caution to readers, I would only add that a bunker syndrome between command centres under pressure is the rule rather than the exception. Command centres quickly adopt personalities. The analogy of personal relationships to those between command centres, with imperfect communications in stressful, threatened and entrepreneurial circumstances, is therefore a useful guide to their likely behaviour. As a redeeming factor in the Falklands Campaign, this was in some senses a creative tension, and it did not prevent our ultimate success.

Command in war involves a matrix of command centres, great and small, but all with mandates (known variously as 'directives', 'orders' or 'rules of engagement') designed to give maximum independence, within specific (often politically-driven) constraints. This is the crucial key to flexibility, and it applies equally in principle to any organisation well prepared for crisis. Operational flexibility, as long as it is founded upon mutual understanding between well-trained command centres, is a fundamental concept for successful military operations. This is almost completely misunderstood in the civilian world, where 'orders' are interpreted as rigidity, rather than as mandates for integrated independence and initiative – thanks partly to the habitual film and television caricatures of the armed forces. A consequence is an exaggerated distinction between commercial and military methods of doing business, which in turn conceals sadly many valuable elements of practical experience each sector has to offer the other.

Creating Super-Togetherness for the Big Push

Fearless' aim had now changed, from being an instrument ready for her operational role, to the discharge of this role in earnest, as required and directed. My task in achieving the remaining relatively modest metamorphosis to a war footing was now mainly a matter of human relations, and education in war imperatives and stratagems. To this I devoted all the powers of leadership accumulated over my career, drawing fully upon the fruit of seeds I had taken so much trouble to sow and let flourish in *Fearless*. From a demanding and pushy Captain who, whilst quite approachable, never seemed satisfied, I found myself a source of strength, dependability and reassurance – and I dare say some inspiration. Correlli Barnett (in lectures 1966 – 2001) has written that 'Leadership is a process by which a single aim and unified action are imparted to the herd. Not surprisingly it is most in evidence in times or circumstances of danger or challenge. Leadership is not imposed like authority. It is actually welcomed and wanted by the led'.

Field Marshall Lord Slim, with Montgomery perhaps the outstanding Army leader of WW II, offers (1956) a similar slant: 'When things are bad, there will come a sudden pause when men will just stop and look at you. No-one will speak; they just look at you and ask for leadership.'

As we progressed south, cumulative pressures on my time drew down upon my wandering habits. I replaced these by increased use of the public address (PA) and CCTV systems. My senior managers all became able presenters on these media. As Captain, I had previously used the PA system only when there was news or a message of major import to convey to the whole ship's company. Its use had in any case presented me with something of an ordeal, since I retained the residue of a once serious stammer. As *Fearless* moved towards war as a pawn in a dynamic and complicated diplomatic situation, I quickly discovered that – over and beyond the hourly World Service bulletins – my people thirsted for homespun news and comment. And it was from me, their captain, that they preferred to hear.

A central leadership task is to promote a clear view of today's reality and expectation in a constructive way. Fast moving circumstances and a deteriorating or increasingly alarming situation can create a reality lag. Of Sir Ernest Shackleton, the celebrated Antarctic leader and (to my pride) a distant cousin, Dr Alexander Macklin (following the loss of the Endurance in the Weddel Sea ice pack) wrote:

> As always with him what had happened had happened. It was in the past and he looked to the future ... Without emotion, melodrama or excitement he said 'Ship and stores have gone - so now we'll go home. (Huntford, 1986).

Henry Kissinger (1982) writes that:

> The most important role of a leader is to take on his shoulder the burden of ambiguity inherent in difficult choices. That accomplished, his subordinates have criteria and can turn to implementation. (p. 531)

We took the morale pulse from day to day. One evening during our three-week pause at Ascension Island, the very same Royal Marine band that had unleashed such carefree enthusiasm but two months previously in the West Indies boarded to give a concert. The event was well attended but subdued. Concerned at first, I concluded correctly that sober stock was being taken of our prospects. We accepted that our likely fate was to sail south to the Falklands (Figure 7.2), to face hazards new to us all.

I was determined there be no jingoism in my ship. When the Argentine cruiser General Belgrano – of comparable size to *Fearless* – was sunk by HMS Conqueror, I briefed on PA hour by hour on the number of her ship's company reported saved. Soon HMS Sheffield was lost also. Thus, as we now moved inexorably towards the theatre of operations, together we cultivated and nurtured by every available means a waxing, realistic and implacable resolve.

Conquering Fear with Information

The last ten years of intimate experience of crisis and emergency management in the civilian world has assured me that many commercial and industrial managers understand all such principles of motivating and empowering the workforce. But only a few companies really practise crisis or emergencies sufficiently. Consider an industrial workforce sheltering during a prolonged chemical or nuclear emergency. Few major-hazard plant senior managers truly comprehend the need for (or, I have to say, have mastered the important basic techniques of) briefing, reassuring and providing material comfort for their people in such circumstances, probably very frightened, and later (because ill-informed) resentful – and thus poor ambassadors when interviewed by the media. The techniques even of using a PA system effectively in emergency are rare, if not totally unknown. Let me illustrate the need. A senior soldier (Macdonald, 1992) wrote of the air attacks on San Carlos following the landings as follows:

> Being on a ship whilst it is being attacked by aircraft was a completely new experience for me…You get a countdown, you can hear the roar of the engines of the jet aircraft, the whoosh of missiles firing and the cackle of small arms and machinegun fire. Then the whole aluminium and steel fabric of the ship – or cell as it seems at that moment – rocks and shakes in combination with the noise of the bombs exploding. You have no understanding of what is going on outside of your immediate area. ... The Captain clearly understood how everyone was feeling. He calmly announced to the ship's company over the ship's loudspeaker system what was going on, both in and around the ship and what the Navy was doing about it. He was delightfully unruffled and his voice was gentle. He had an immediate, positive, calming effect on everyone, although no one said very much. 'Brits' are conditioned to behave in a quiet self-composed manner in situations of extreme danger no matter what they are personally feeling inside.

Figure 7.2 Map of the Falklands Island

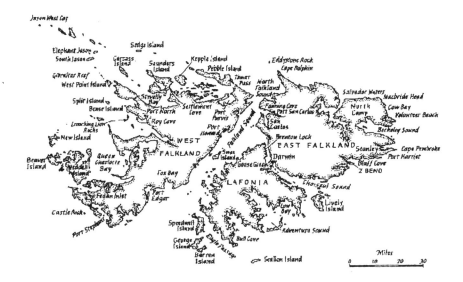

Falkland Sound separates the two main islands, and San Carols Water is the inlet just inside the narrows at the north letting into East Falkland.

The distance from San Carlos to Port Stanley is about 50 miles over extremely difficult terrain. The level ground is generally passable only to tracked vehicles, but even they are obstructed by many rock runs from the mountains, not shown on this map. Movement on foot with action loads is immensely demanding – a huge tribute to the marines and paras who 'yomped' it. (Map: Ewen Southby-Tailyour, OBE, Royal Marines.

Max Hastings covering this phrase wrote (1983):

All over the ship, more than a thousand men, many of whom never saw the light of natural day, strained to catch his voice over the hum of vents and air conditioners, the roar of machinery and the constant traffic of transmissions, telephones and hurrying figures.

He added (2000):

In the wardroom of *Fearless* a scattering of spare pilots and survivors of lost ships clustered round the loudspeaker in their life jackets and respirators.

Battle of Attrition at San Carlos – Operational Decisions

Space does not permit me to describe the many fascinating considerations governing the final approach of the Amphibious Force to the assault, nor the unfolding of events. All this has been covered expertly in a range of fine books, including those by Max Hastings (2000; Simon Jenkins covered the political elements in their excellent early joint book, 1983), Sandy Woodward (1992), Michael Clapp (1993), Julian Thompson (1992), Nick Vaux (1986) and Ewen Southby-Tailyour (1993). My own contributions at this juncture majored in our anti-submarine measures and the crafting of a deceptive approach to the Falkland archipelago from the northwest. The aim of the latter plan was to indicate for as long as possible that we were heading for the East Coast somewhere around Port Stanley – unnecessary in the event, since the Amphibious Force was not detected. Michael Clapp delegated to me the tactical handling of the Amphibious Force during the critical final 24-hour transit to San Carlos. To lead, finally, the long line of ships into the Falklands Sound, keeping very close to bluffs of West Falkland Island in total darkness and hence clear of the central channel which was where the Argentines might have laid mines, was certainly amongst the proudest episodes of my life.

Immediately following the landings at San Carlos during the early hours of 21 May 1982, it fell largely to me on Michael Clapp's behalf to organise the stationing of ships in the anchorage. Here the recent Norwegian fjord experience with camouflage proved valuable. We anchored ships packed together around the relatively slight major land features in a way that would inconvenience the Argentine fast-jet weapon-launch flight paths. Flaws in my initial calculations led to hits on two logistic ships, the bombs happily failing to explode. Thereafter no strikes on troop – or cargo-carrying ships in the anchorage were achieved. In case Argentine daylight intelligence of this unconventional arrangement might encourage high-level night attacks, I had all ships moved to dispersed positions at nightfall, resuming the compact formation before dawn.

I had also to decide from where to fight *Fearless*. My small 1960-vintage Operations Room was filled entirely by Michael Clapp and his staff. Moreover, my ancient air-defence radar saw nothing beyond the surrounding hills; from amongst which attacking aircraft would emerge some ten seconds from target (e.g. *Fearless*). Having endured the first air attack as a frustrated supernumerary without even a radar monitor at my disposal, I decided to migrate to the missile/gun-direction platform at the very top of the ship, a level above the navigational bridge. This was better. From here I could see incoming aircraft as soon as anyone else. I could direct the manoeuvring of the ship to present the least target and best weapon arcs to the Argentines' favourite flight paths. I had first-hand observation of each attack on the anchorage and its results, and could provide a deputy's eyes for my Commodore. I could moreover set a direct example to my people, some fifty of whom were on the upper deck manning weapons from missile launchers to infantry automatic weapons – not to speak of Michael Rose and his SAS team, who emerged from their portacabin during attacks to join the shoot.

Major Ewen Southby-Tailyour (1993) wrote:

> He was sitting in the Captain's bridge chair obviously resting from air attacks. Modern naval warfare requires ships to be fought by electronics in an operations room and not from first-hand knowledge from the bridge – or, in Jeremy's case, the gun direction platform. But *Fearless* was not a conventional ship with sophisticated defences ....

One attack splattered us with shrapnel, I believe from a shell fired by a friendly frigate. There was a yelp as a sailor beside me was clipped, and louder cries below announced injuries amongst the young Bofors guns' crews on the bridge wings. I looked over to see men leaving their positions to tend the wounded. Having anticipated this comradely reflex, I used all my available decibels to get the guns back into action in time for the next passing predator. Stretcher teams succoured the casualties.

Selection of this command position had its amusing side. It was a new experience to stand unprotected in the path of fast jets bent on one's destruction. Adrenaline and an imprecise feel for our life expectancy carried us through the first day in almost carnival spirits, typical of novices to real war. It gradually became clear that prospects of survival were quite good. With just a few exceptions, we had substantial warning of attacks. I would update the ship's company on PA before moving to my command platform, whereupon the Navigation Officer John Prime took up the commentary from the bridge. My arrival 'up top' signalled further to the weapons-direction teams that action was imminent. As the days went by, I sensed wryly a growing inner tardiness to indulge this routine exposure – although I never failed to appear except for one of the attacks that surprised us!

Command Pacing

Certainly the issue was not straightforward. Almost every night, during some fourteen hours of total darkness, *Fearless* deployed from the anchorage. The aim might be a special forces insertion, the meeting and escorting of the submarine *HMS Onyx* into the anchorage for a pit-stop and out again a few hours later, or the escorting of a convoy outbound, to rendezvous with another to be escorted back. The rule was to be within the relatively safe confines of the defended San Carlos anchorage by the first glimmerings of daylight – even if on two notable and nerve-racking occasions we did not achieve this. The two splendid Johns, Kelly and Prime, took turns at conducting the ship during part of these night-time forays once we were clear of Falkland Sound while I cat-napped on my sea-cabin couch, microphone in hand. Long years had taught me to absorb a question in my sleep and wake to provide considered response. This was an operation of serial marathon dimensions. We not only survived, but sustained our edge. With beguiling delusions of immortality, I felt we could continue indefinitely – wrong of course. We did between us manage the 100 days, with what seemed capacity to spare. I recognised with hindsight that there was a long-term price to pay, manifest within my family.

Night Operations – Issues of Risk and Reverse Critical Path Planning

I have selected one operation of particular interest to illustrate some commonplace command problems. The central issue was what I would term 'reverse critical path planning'. Following the loss of the Cunard container ship Atlantic Conveyor with much of the force's military helicopter lift, the battle-fit battalions of Commandos and Paratroopers moved forward some 50 miles to invest the northern approaches to Port Stanley mainly by the celebrated and heroic expedient of 'yomping' heavily laden across the appalling Falklands terrain (Thompson, 1992; Vaux, 1986). Formations of the reinforcement brigade, who were to take the southern flank, were not equally robust in pure fitness terms, and anyway time pressed. It thus became necessary to take them forward by sea round the southern extremities of the islands.

Intrepid took the Scots Guards, launching them by landing craft some considerable distance short of their destination. They survived harrowing hours in very bad weather, amongst other nasty hazards, under Ewen Southby-Tailyour's redoubtable command (Clapp & Southby-Tailyour, 1993, Southby-Tailyour, 1993; Woodward, 1992). Next day I persuaded Michael Clapp to let *Fearless* take the Welsh Guards that night, notwithstanding the serious disruption to command communications this would cause for some hours. The plan was to take two of my own landing craft (two had to be left in San Carlos), rendezvous at sea with *Intrepid*'s four off the entrance to Choiseul Sound, send four of the six forward with the Welsh Guards embarked to Bluff Cove and bring two back to San Carlos. I was determined to get nearer to their destination (Bluff Cove/Port

Fitzroy) this time, whilst avoiding the footprint of a land-based Exocet site the Argentines were believed to have established south of Port Stanley. I had to be back at San Carlos by first light. This all involved some nice navigational calculations.

We sailed in nasty bumpy weather and proceeded south through the Falkland Sound and Eagle Passage, then northeast. We reached the rendezvous in good time, calm sea and comparatively docile weather. For reasons I need not recount but then unbeknownst to me (Southby-Tailyour, 1993), *Intrepid*'s landing craft had been unable to sail to meet us. Against a calculated reverse critical path, I waited longer than I should have dared (continuous RPD, and perhaps a bit of creative decision making too), finally launching my two landing craft with half the Welsh Guards and legging it for San Carlos. Twilight then full daylight found us still forcing our way back up the Falklands Sound. The Argentine Air Force did not call our bluff.

It was decided that *Fearless* could not again be released on such an extended and tenuous communications tether from the Land Force, and that she was anyway too crucial a command asset for such high-risk exposure. To underline the point, one of the two landing craft then launched, Foxtrot Four, never returned. She was destroyed two days later with six of her crew in a savage and pyrrhic air strike, three of the four Argentine aircraft being themselves destroyed with their pilots by a Sea Harrier counter-attack within one minute – such is the pace of modern warfare.

The next night *Fearless* once more escorted a convoy outbound and collected another inbound from the northern ocean holding areas. Some complications left us with another high-risk late return to San Carlos. There was an assessed threat from land-based Exocet on Pebble Island to the west as well as air attack, and once again our luck held. Meanwhile the less lucky remainder of the Welsh Guards embarked in the logistic ship Sir Galahad, and moved forward to their tragic destiny.

Victory

A week later it was all over and *Fearless* was anchored in Stanley Harbour. Onboard as my 'guest' was the Argentine late Malvinos Commander, General Mario Menendez. We had some interesting discussions. The General had a final advantage however; he kept 'Spanish' hours with siesta to the local clock, whilst I was working a 20-hour day based on UK time (four hour ahead of the Falklands)! He poured the whiskey, with which I had had his cabinet stocked.

Principal Lessons

1.   Successful command and leadership have their foundation in thorough preparation:

- Know your team. Win the trust of your people, and keep faith with them – above all in hard times.
- Within reason, be prepared to do anything required of your people and demonstrate this from time to time.
- Know your company/equipment/site as intimately as practicable – its capabilities and its weaknesses.
- Set the right strategy and follow it in everything that you do.
- Think through every conceivable credible development; you may still be surprised, but you can then deal with this from a position of well-founded confidence.
- Train and prepare to a high standard – and prepare your team as well as yourself for the unexpected.

Then when it all starts:

2.   Set your aim and objectives, and review them repeatedly.
3.   Ensure your resources are sufficient for the task and are applied efficiently.
4.   Define the mandates of command centres with care; recognise and manage the 'personality' that command centres will develop; whilst giving direction (the 'what'), let them get on with their work (the 'how') with as little interference as possible.
5.   Win the information high ground, retain it and exploit it. It is essential always to know the present situation and how it is developing. Without accurate and timely processed high-quality information you are lost.
6.   Based upon sound strategy, thorough analytic preparations and exhaustive 'what-iffing', naturalistic ('recognition-primed') decision making can, with discrimination, sensibly be applied beyond the envelope of your previous direct experience. This prepares the ground for creative decision making also, as best may be.
7.   Lead vigorously without fear or favour, and manage morale; recognise and respond to human needs – fear, anxiety – and do everything necessary to build, sustain and nurture this; be as upbeat as the situation permits.

8. Be yourself, but you need to practise the 'art of acting' within this envelope of credibility; develop and use, therefore, all the modulation of delivery and style (demeanour, tone, volume and tempo) that you find useful.
9. Guard against the very real dangers of over-confidence and the corrupting effect of power.
10. Provide your team with the best information you can, not least as an antidote to fear.
11. Be constantly aware of the time-line.
12. Pace yourself, conserve stamina, delegate and prioritise your cadence of command.

The reader will recognise the importance of the leadership element in any endeavour of the nature I have described. As a final reference, I therefore recommend *Shackleton's Way* by Margot Morrell and Stephanie Capparell just published (2001). This book is an examination of Sir Ernest Shackleton's career as an inspiration and guide for any leader, to which I have been privileged to contribute.

References

Barnett, C. (1966). Lecture on leadership. First delivered to Army Staff College, Camberley, and currently to Cambridge University Engineering Department's Manufacturing Leaders Programme.

Churchill, W. (1933). *Marlborough, his Life and Times.* London: Harrap.

Clapp, M. & Southby-Tailyour, E. (1993). *Amphibious Assault Falklands, the Battle of San Carlos Water.* Barnsley, Yorkshire: Leo Cooper, Pen and Sword Books.

Clark, A. (1994). *Diaries.* London: Phoenix.

Clausewitz, V., K. (1832). *On War.* Translated J.J. Graham (1908) and edited by Anatol Rapaport (1968). Harmondsworth: Pelican.

de Guingand, F. (1947). *Operation Victory.* London: Hodder & Stoughton.

Financial Times (1 Aug 1998). FT Inside Track. London: FT.

Flin, R. (1996). *Sitting in the Hot Seat. Leaders and Teams for Critical Incident Management.* Chichester: Wiley.

Gordon, A. (1996). *The Rules of the Game – Jutland and British Naval Command.* London: John Murray.

Hastings, M. (1984). *Overlord D-Day, June 6th 1994.* New York: Simon & Schuster.

Hastings, M. (2000). *Going to the Wars.* London: Macmillan.

Hastings, M. & Jenkins, S. (1983). *The Battle for the Falklands.* London: Michael Joseph.

Howarth, S. (1998). London Financial Times (1st August), Inside Track.

Huntford, R. (1986). *Shackleton* (quote from Macklin, A H, 'Shackleton as I knew him', unpublished manuscript). New York: Atheneum.

Kissinger, H. (1982). *Years of Upheaval* Boston: Little Brown.

Klein, G. (1997). The recognition-primed decision (RPD) model: Looking back looking forward. In C. Zsambok & G. Klein (Eds.) *Naturalistic Decision Making*. Hillsdale, NJ: Lawrence Erlbaum.

Klein, G. (1998). *Sources of Power. How People Make Decisions*. Cambridge, Mass: MIT Press.

Lane Fox, R. (1973). *Alexander the Great*. London: Allen Lane.

Lane, R. (2000). The fog of war, a personal experience of leadership. In C. McCann & R. Pigeau (Eds.) *The Human in Command. Exploring the Modern Military Experience*. New York: Kluwer/Plenum.

Larken, E.S.J. (1992). Command as a feature of crisis prevention and crisis management in the offshore industry. In *Proceedings of the Offshore Safety: Protection of Life and the Environment conference*. The Institute of Marine Engineers in association with The Royal Institute of Naval Architects, London.

Larken, E.S.J. (1995a). Practical Emergency Management. In *Proceedings of Major Hazards Onshore and Offshore Conference II*. The Institute of Chemical Engineers Symposium Series No.139, Rugby.

Larken, E.S.J. (1995b). Practical Emergency Management. In *Proceedings of the Emergency Planning and Management Conference*. Institute of Mechanical Engineers Transactions 1995 – 7, London.

Leach, Admiral of the Fleet, Sir Henry (1993). *Endure No Makeshifts – some Naval Recollections*. Barnsley, Yorkshire: Leo Cooper, Pen and Sword Books.

Longford, E. (1969). *Wellington – the Years of the Sword*. London: Weidenfeld & Nicolson.

Macdonald, R. (1992). *The Heart of Leadership*, Defence Fellowship, Ministry of Defence (UK), London; University of Southern California.

Manchester, W. (1978). *American Caesar – Douglas MacArthur 1880-1964*. Boston: Little Brown.

McCann, C., & Pigeau, R. (2000). (Eds.) *The Human in Command: Exploring Modern Military Experience*, New York: Kluwer Academic / Plenum.

Montgomery, B. (1958). *Memoirs*. London: Collins.

Morrell, M. & Capparell, S. (2001). *Shackleton's Way*. London: Nicholas Brealey.

Orasanu, J. (1997). Stress and naturalistic decision making: Strengthening the weak links. In R. Flin, E. Salas, M. Strub & L. Martin (Eds.) *Decision Making Under Stress*. Aldershot: Ashgate.

Orasanu, J. & Fischer, U. (1997). Finding decisions in natural environments. The view from the cockpit. In C. Zsambok & G. Klein (Eds.) *Naturalistic Decision Making*. Mahwah, NJ: Lawrence Erlbaum.

Rose, M. (1998). *Fighting for Peace*. London: Harvill Press.

Royal Navy (1961). *Conduct of Submarine Operations*. Flag Officer Submarines publication. Gosport, Kent: Royal Navy.

Royal Navy (1995). *The Fundamentals of British Maritime Doctrine*. London: HMSO.

Slim, W. (1956). *Lecture to US Army Military Academy, West Point*.

Southby-Tailyour, E. (1993). *Reasons in Writing*. Barnsley, Yorkshire: Leo Cooper, Pen and Sword Books.

Strutt, J., Allsopp, K., Lyons, M., Larken J., & Værness, R. (1997). Development of models and data for quantification of human reliability in emergency management. In *Proceedings of the Risk Assessment of Offshore Installations Conference*. ERA Technology, Church House Conference Centre, London.

Terraine, J. (1989). *Business in Great Waters – the U-Boat Wars 1916-1945*. London: Leo Cooper.

Thompson, J. (1992). *No Picnic*, (2nd ed.) Barnsley, Yorkshire: Leo Cooper, Pen and Sword Books.

Turnbull Report (1999*). Internal Control – Guidance for Directors on the Combined Code.* London: The Institute of Chartered Accountants.

Vaux, N. (1986). *March to the South Atlantic.* London: Buchan & Enright.

Woodward, J. (1992). *One Hundred Days; The Memoirs of the Falklands Battle Group Commander* (quote from forward by the Rt. Hon. Margaret Thatcher). London: Harper Collins.

8 Military Commander – Royal Marines

Andrew Keeling
Berwick Associates

Lieutenant Colonel Andrew Keeling was the Commanding Officer, 45 Commando Royal Marines from 1985–1987. By then he had been commissioned for 23 years and had served in a number of trouble spots around the world; Borneo and Aden in the 60s, Cyprus in the 70's and the Falklands in 1982. He was something of an expert on Amphibious Warfare. He had instructed at Dartmouth Naval College and at Sandhurst Military Academy; he was a graduate of three separate military Staff Colleges. In short he had a wide range of relevant military experience. He felt comprehensively prepared for his new, high profile appointment as CO of 45 Commando, except that he had never served in Northern Ireland.

Northern Ireland – 1986

My total lack of Northern Ireland experience would not have mattered but for the fact that at the same time as I took command of 45 Commando in April 1985 we were told we were to do an Emergency Tour in West Belfast from 2 July – 10 November 1986.

How did I feel on becoming a Commanding Officer?

- Professionally satisfied. This was the best job I could have at this stage of my career, and maybe ever.
- Generally well prepared by virtue of my training and experience.

- Keen to embark on a range of comprehensive training exercises to impress my own stamp on my team and to ensure that we were all prepared for whatever would come our way.
- Somewhat apprehensive about the forthcoming Belfast tour and very conscious that in the forthcoming 14 months I had a lot of work ahead of me to prepare for it.
- Aware that my men knew that I was a Northern Ireland 'novice', and unsure what they would think about that.
- Conscious that I was going to spend most of the next 2 years separated from my wife and teenage children – not something that I relished.

Background – Setting the Scene

By the mid-1980s 'The Troubles' in Northern Ireland had been raging for over 15 years. 45 Commando had already done 9 Emergency Tours in the Province, 8 of which had been in Belfast. Many of my 700 men and women had served there before. Some had done so several times, either with 45 or with other units. By 1985 there was a considerable military folklore surrounding service in Northern Ireland; special clothing and equipment, special tactics, lots of jargon, a whole lexicon of names (streets, police stations, pubs, 'players', community leaders and so on) as well as an endless supply of stories about famous and less well-known incidents that had (or had not!) happened over the years. Many servicemen, including Royal Marines, saw themselves as specialists in Northern Ireland. They had served there many times, were fluent in the jargon, had very firmly ticked the Northern Ireland box and had drawers full of green tee-shirts. I had several of these very experienced people in my unit, and many more with just a tour or two under their belts. Yet as the leader of this proud group of men I was the only senior officer in 45 Commando with no Northern Ireland operational experience at all and was very aware of this. I knew some of the stories and could, rather clumsily, handle some of the jargon, but not surprisingly I felt like a novice.

Preparation – The Vital Stage

However, I had a year with the unit before our tour was to start. We had a full programme of exercises and other commitments on which to concentrate, including 3 months Arctic training in North Norway early in 1986. These various activities all provided me with an excellent opportunity to manoeuvre myself, my command team and indeed the whole unit into a position of competence and confidence about our ability to embark on a successful tour. So I set out to use the year to:

- Get to know my people as well as possible, particularly the command team.
- Form balanced teams at all levels, including my own command team.
- Ensure that, whatever the threat, the chain of command was sound, robust and flexible.
- Work us all hard and really put us through our paces. Royal Marines respond well to being extended, as long as they can see that it is relevant and sensible.
- Give praise where it was due.
- Enjoy ourselves as much as possible, both professionally and socially.

Leadership – Delegation – Teamwork

I used every opportunity I could to delegate to subordinates and made it clear that I expected this to ripple on down the chain of command. Groups of men were encouraged to get away on their own, either for military or adventure training. Such activities provide rich opportunities to bind men together and to generate healthy respect between ranks. Whether or not they realised it at the time I don't know, but as the year wore on they were getting to know each other very well, and to trust each other and have confidence in each other, sometimes in formidable circumstances. We became serious about our sporting efforts and won a number of competitions: I wanted everyone to experience the sweet taste of success and to understand that there are no prizes for coming second. All of this called for positive and active junior leadership, of course, and resulted in a lot of young men enjoying some challenging experiences together.

At about this time I came across a definition of leadership from an anonymous source that seemed particularly appropriate:

> Good leadership is not composed of social goals, quotas of talent or political ambition and designs. It is an earthy binding of leader and warrior in common cause, and precisely to the degree that bonding is loosened for whatever reason, leadership, performance and the mission suffer.

I worked hard on the 'earthy binding' and the 'common cause'. I also remembered my military hero, Joseph Lawrence Chamberlain, Colonel of the 20th Regiment of Infantry, Maine Volunteers at the Battle of Gettysburg, who is reputed to have believed that there are just two things an officer must do to lead men: 'You must care for your men's welfare.

You must show physical courage' (Shaara, 1974). An over-simplification perhaps, but none-the-less a powerful message.

As the end of 1985 approached, and with the very concentrated 3-month training period in North Norway looming, I began to realise that, come Easter the following year, we were going to be eyes down on preparation for Belfast. Convinced of the need to be proactive rather than reactive whenever possible, I took two particular decisions that were to pay off handsomely:

- I made my second-in-command (2IC), an extremely able officer, responsible for preparing the unit in all respects for the Northern Ireland training period which we would embark on in April. This freed me up to concentrate on the job in hand.
- I arranged to spend a few days with the CO of the West Belfast battalion in situ late in 1985. Every CO deploying to Northern Ireland was entitled to a recce, with a handful of his key staff, about 3 months before deploying. But I was keen to start my personal education programme before that. It proved invaluable for I learned a lot, and everything I learned was current, not folklore, which put me in a strong position with my own people, many of whom *thought* they knew the form, but actually were way out of date. This was important, for as I was discovering the Northern Ireland military scene was distinctly dynamic, owing to politics (the Anglo-Irish Accord et al.), technology, the steadily increasing skill of the terrorist and the gradually changing mood of the people. I realised with relief that, although a novice, I was completely 'folklore-free' and therefore to some extent liberated from the limiting effects of the folklore. This was to be a considerable personal asset as we got into the detailed preparation.

Training – General Conditioning

Although the North Norway environment was in all respects very different from Belfast, our Arctic training gave us a marvellous opportunity to prepare for the tour to come. We were living quite rough in company bases, more or less as we would in Belfast, and for 3 months I was cheek by jowl with my own key headquarters staff. Having made my requirements and expectations clear to my subordinates I delegated as much as I could. Company commanders ran their own camps and their own programmes in accordance with my Directive. The training was hard and as realistic as possible; it isn't difficult to generate challenging situations in winter in the mountains of North Norway! My logistics team too had a demanding task, supporting a widely dispersed unit with poor communications and in a very harsh climate. They participated fully in all our exercises as well as maintaining our camps, the vehicle fleet and so on.

Each week I got the command team together for a good long chat, usually followed by a meal together. We changed the location of these 'O Groups' with each company taking a turn to host them. We were to repeat this pattern in Belfast, although there we ensured that with these weekly meetings, as with everything else we did, there was no pattern. And in addition to the winter training, which is always an exhausting experience, the key people in my headquarters team were increasingly involved in a whole raft of preparations for the Northern Ireland training package and for the tour itself.

After Norway we all went home for our Easter Leave – the last proper break we were to get until November. We knew we had a hard 6 months ahead of us.

Training – Specific for the Task Ahead

By 1986 countless battalions, like mine, had gone over the water for emergency tours. To prepare us the Army had set up a special Northern Ireland Training Team (NITAT), and had designed a detailed series of specialist NITAT courses. At the start the 2-month package looked somewhat daunting, but in the event it was planned and delivered with great care and with consummate professionalism. The NITAT staff knew their stuff very well indeed and were first class instructors. They prepared us well.

The first of numerous courses was 'The Commanders' Course'. About 60 officers, warrant officers and senior NCOs went for a week to an Army camp in Kent for a detailed orientation on the Province in general and West Belfast in particular. We covered history, politics, the mood of the moment (as far as anyone could judge it), the RUC, the intelligence set-up, the threat, the likely pattern of operations, tactics (in great detail), special equipment and so on. We learned about 'police primacy', what 'an acceptable level of violence' meant and that we were to adopt an attitude of 'relaxed alertness'. This last one was hard to sell to some of my men. We also learned about the rest of the Army in Northern Ireland – all new stuff to me. Who would we be working alongside, who was to be my boss and so on. It was very comprehensive. Like everyone else I was a student, and for once in my life I tried hard to be a very good one. My only personal contribution to the week was to insist that we all met every morning at 6 a.m. for a PT session before the day's work began – and the first morning I led the run and physical jerks myself. I thought this would be good for the 'earthy binding' process, and I'm sure it was.

The Tour Starts

Days turned into weeks and before I knew it we were there. I took over from the CO of the Welsh Guards at Springfield Road Police Station at 10pm on 2 July, two days before my 43rd birthday. This antiquated building, deep in the heart of nationalist, Republican, Catholic West Belfast was to be my home, and that of about 50 of my people, for the next 20 weeks or so. It was a rabbit warren of a place, crammed between houses and shops and heavily fortified. Like all the other locations living conditions were extremely cramped and very basic.

Springfield Road was apparently the most attacked police station in the Province. It was also the Headquarters of 'B' Division of the RUC, and housed the office of 'BD 1', the RUC Chief Superintendent who was responsible for West Belfast. I soon came to have great respect for this fine officer, and for his team, and was extremely grateful to be co-located with him. My three company commanders each had their own base, all within a few miles of my HQ, and my logisticians were based a little further away near the Musgrave Park Hospital. Each company commander had his own patch, each of which was approximately a third of my whole area.

Our Mission was quite simply, 'To support the RUC in defeating terrorism'. Straightforward enough in a way, but this was an extremely complex situation for all sorts of reasons, and the terrorists were some of the most skilled in the world at the time. In terms of street-violence, shootings, bombings, ambushes, highjacks, house take-overs and all the other crazy antics of terrorist-enflamed inter-community hatred, the 80's was a relatively quiet period. The long-suffering people of West Belfast were becoming increasingly fed up with their lot as the jam in the sandwich of apparently endless violence, death, destruction and social upheaval. But this is not to say that they regarded us 'Brits' as their friends. Far from it, the residents of West Belfast made it very clear that we were extremely unwelcome. Most of them probably genuinely hated us; they certainly behaved as if they did. I found it a very gloomy place.

In supporting the RUC we set out to re-assure the population as a whole, to deter aggression and, when necessary, to combat active terrorism. We carried out literally thousands of patrols, each one of which was in support of the RUC. Every one had to be planned meticulously and executed with great care and imagination. Every pair of policemen on the beat was accompanied by at least 16 armed Marines who worked in teams of 4 and wove their way in and out of the streets around them as they went about their business. When possible they were supported by a surveillance helicopter. My men were frequently subjected to abuse, both verbal and physical, and they had to show restraint at all times. On average we all worked at least 16 hours a day, 7 days a week.

My Role as Commander

So how did I manage my time, as the unwelcome military squire of a sizeable but pretty inhospitable patch (with a population of about 90,000) of a British city? As already implied my area was divided into sub-areas, each commanded by a mature, experienced, well trained and certainly very well trusted (by me) subordinate. Although I occasionally set up and commanded a unit-level large-scale operation, this was the exception and probably only happened 4 or 5 times. How could I exercise useful and effective command? With hindsight it seems surprising that there was no formal advice on this subject. Rather it was down to each individual CO to do it his way – very much the ethos of British military command at that time. As an aside, and although I was not aware of this at the time, the revolution in military doctrine was about to happen, and 'Mission Command' would help future commanders (including me) greatly in this respect.

I decided that I had 3 principal command responsibilities:

- To establish and maintain a sound working relationship with the RUC. We were, after all, in Belfast to 'support the RUC in defeating terrorism'.
- To be an effective leader to my Commando.
- To establish a good relationship with and gain the respect of my newly acquired superior headquarters, 39 Infantry Brigade in Lisburn.

The RUC was, in a sense, easy. For a start they desperately needed our help. On the other hand they had a reputation for being quite difficult, for as their military muscle came and went over the years they inevitably found some units more to their liking than others. I determined to get to know BD1 well, early on. He, in turn, welcomed me warmly although made it very clear that my unit, no stranger to Belfast in the bad old days, had a certain reputation for hasty and sometimes inappropriate action that would certainly be extremely unhelpful in the present climate. I took his warning to heart. I made it my business to see him informally very regularly, to attend his weekly meetings myself and to get to know his command team too. I also took every opportunity I could to talk to and identify with his officers on patrol, at incidents and so on. I can also remember playing snooker with them – and winning! The only concession I felt I had to make was over my best intentions to remain 'dry' for the whole tour. But I quickly realised that this threatened to underpin my novice status, for I had not appreciated the significance of the Bushmills bottle behind closed police doors!

Leading my Commando in these circumstances required thought and a new personal approach too. Although there was plenty of paperwork to be done, I felt strongly that time spent in my office was time almost completely wasted. I quickly realised that I needed to spend as much time as possible out on the ground with my people as they 'supported the RUC'. This enabled me to gain a sound understanding of what was going on and of the stresses and strains that they faced on a daily basis. I visited each company base nearly every day and certainly felt as if I was just part of the furniture rather than being a visitor that somebody had to look after. In doing this I spent a lot of time driving around the patch, thus becoming very familiar with the geography. Sometimes I would hop out of my vehicle and join a patrol for a while – they seemed to like this and the RUC certainly found it quite novel. My almost constant mobility meant that my communications had to be very good – I spent 20 weeks with a radio plug in my ear and was frequently on the air.

Building a sound relationship with the Brigade Commander and his staff, whom I had never met before, was also important. I met them at Lisburn for their fortnightly meetings, and occasionally in between when one or more of them came to see us in Belfast. My main contribution was to ensure that my staff and I kept them fully informed at all times. It was important that this included the bad news as well as the good news.

Managing Major Incidents

Routine life was peppered with major incidents of various sorts, all of which to some extent required a more static presence on the streets than our routine, mobile patrolling. Static troops make easy targets – a point not missed by Irish terrorists over the years. These major incidents were often complex and demanding in terms of command and control. They also invariably attracted the attention of a range of other police, military and media agencies, all of whom quite naturally homed in on the Incident Command Point (ICP). Major incidents also inevitably reflected, and sometimes generated, a heightening of tension. They included riots, bombings, shootings, searches, marches, demonstrations and a major funeral or two. Sometimes incidents were hoaxes, and were actually attempts by the terrorists to generate a 'come-on' where troops would stumble into an ambush of one sort or another. I am pleased to say that we never fell into this trap.

Apart from the handful of unit-level events that I commanded myself, major incidents were run on the ground by the relevant company commander from a hastily established mobile ICP. I always made it my business to get to the ICP as quickly as possible to support him. I tried to do this by doing whatever was necessary to allow the company commander to give his full attention to the incident. This usually included looking after the inevitable crowd of agencies, goofers and other interested parties who often seemed to think they deserved the full attention of the commander on the ground.

I was always very aware of the need to avoid getting in the company commander's way and to allow him to 'run' the incident. However, I found that quite often he was glad of the opportunity to ask my advice directly and to chat informally through the problem. It was also invaluable to me to be there on the ground, seeing for myself, feeling what was happening, smelling the atmosphere, and experiencing the tension and the pressures on my men. When you are dealing with dangerous situations like this out on the street there is so much to be aware of to gain anything like a full picture. This undoubtedly includes such things as the mood of the people around you (both hostile and friendly), the attitude of the agencies with whom you are working and a whole range of local knowledge items. I always much preferred to be there at the scene, albeit to some extent in detached mode, than closeted in my Operations Room back at my Headquarters where, with the best will in the world, my own real understanding of what was happening on the ground would have been significantly limited. What it came down to, quite simply, was that the best place for me to exercise effective command was at, or close to, the epicentre of the incident.

I also found it useful to be able to keep my own headquarters, and sometimes Brigade headquarters too, fully informed about what was happening on the ground, and more importantly what I thought about it. It meant that I could relate my own decision-making to exactly what was current, and that I could communicate this readily to those above and below me in the chain of command. It often meant that, in a way, I was acting as the company commander's rear reporting link. But it worked for me, and my boss at Lisburn (15 miles away) certainly felt re-assured to have my immediate input on the spot over delicate and potentially high profile issues. Meanwhile my 2IC always ran the unit Operations Room during an incident, thus freeing me up to be out on the ground with my people at the scene.

Each major incident was a unique event and they always brought their own risks and challenges. Although occasionally set-piece (for instance, in response to a planned march or a terrorist's funeral) most were totally unexpected, and whether set-piece or not they always generated a range of unpredictable threats, options and responses. They were always potentially life-threatening events, as we inevitably had a large number of men out on the streets, often in rather predictable places, and all of them were potential targets for terrorist bombers or snipers. The key to commanding these incidents, I decided, was:

- To be very clear about the Mission, or the Aim.
- To issue very clear and simple instructions to subordinates.
- To locate myself where I was best able to exercise effective command.
- To delegate, and then allow subordinates to get on with their tasks unmolested.
- To ensure excellent communications at all times, and to keep the chain of command well informed.
- To be, and to be seen to be amongst my men as they did their best to handle these dangerous and often very trying incidents.
- To be as supportive as possible to my people.

What Makes Success?

45 Commando's tour in West Belfast was judged to be particularly successful. Part of this was because of our tangible results – our tally of successful operations. But in reality most of it was down to the attitude of my men on the streets. In spite of unremitting provocation, and determined attempts by some highly intelligent and hard-core (although luckily for us, not very brave) terrorists, they maintained their standards of professionalism, decency and 'relaxed alertness' to the end. I like to think that I made my contribution to their efforts by overseeing their thorough preparation, by keeping close tabs on the pattern of their daily lives and by ensuring that the command chain which supported them and enabled them to function properly was working for them all the time. But ultimately it was down to them; it always is.

How did I Feel at the End of our Northern Ireland Tour?

- Above all hugely relieved and satisfied that we all went home, relatively unscathed.
- Very tired. The responsibility of commanding over 600 men for a protracted period in an extremely sophisticated and hostile urban counter-terrorist environment was undoubtedly stressful and draining.
- Delighted and very proud that my men had consistently demonstrated their sheer professionalism, flexibility, patience and good humour.
- Grateful for the training that we had received from NITAT, and the support we had had enjoyed from a wide range of RUC and military units and agencies, none of whom we had known before the tour.
- Pleased that I was no longer a novice!

Operation 'Haven' – North Iraq, 1991

Andrew Keeling, by now a Brigadier, took command of 3 Commando Brigade Royal Marines, a force of about 5000, in April 1990. As a light force they were inappropriate for the armoured operations of the Gulf War, although they were designated as a reserve brigade during the war. During the Coalition's protracted air operations in January and February 1991 they were carrying out their normal winter training in North Norway. Throughout this period he felt that it was very likely that after the inevitable land war he and his brigade would be sent to clean up Kuwait. He was wrong!

Background – Setting the Scene

The Kurds say themselves that they 'have no friends but the mountains'. Throughout history they have been denied the right to enjoy their own sovereign state. They probably came closest to achieving their dream of an independent Kurdistan in the aftermath of the First World War following the collapse of the Ottoman Empire. The Treaty of Sèvres (10 August 1920) created the legal where-with-all for this to happen, but it was never ratified. Three years later The Treaty of Lausanne bowed to the demands of the new Turkish Government, and yet again the Kurds were stateless. The British Government had strongly supported the cause of the Kurds; at Lausanne Lord Curzon (the British Foreign Secretary) had told the Turkish representative unequivocally that 'The whole of our information shows that the Kurds, with their own independent history, customs, manners and character, ought to be an autonomous race', but to no avail.[1] (See Bulloch & Morris, 1992; McDowall, 1996 for further details).

For 4,000 years the Kurds' homeland has been the Zagros Mountains that straddle the border regions of the countries we now know as Turkey, Syria, Iraq and Iran. Estimates of the Kurdish population vary between 15 and 25 million. What is certain is that none of the 'host' countries mentioned above have a high regard for Kurds. Kurds have always been generally regarded as troublesome. There is probably a good reason for this. They are a proud people, very aware of their deep cultural links and of their geographical stability over thousands of years. They are also fiercely tribal and their history is littered with costly internal disputes that have probably created far more problems than they have solved.

After 1968 Iraqi Kurds (numbering about 3 million) had endured particular hardships from Saddam Hussein's brutal Ba'athist regime. Northern Iraq had, in effect, been a war zone for much of this period as battles and campaigns between Government forces and Kurdish 'peshmergas' (irregular soldiers – the word 'peshmerga' means 'those who are prepared to die') ebbed and flowed across the beautiful mountainous landscape to the east and north of the Kurdish cities of Kirkuk and Mosul. Kurds had occasionally enjoyed local successes, but seldom more than that. During the 70s and 80s they suffered mass deportations, the destruction of literally thousands of villages and in 1988 widespread chemical weapon attacks. Many thousands of Kurds were killed or otherwise disposed of by Saddam's horrific regime.

In April 1991 I was supremely unaware of all of this. Of course I had followed the fortunes of the Gulf War meticulously, courtesy of Sky and CNN – after all, I fully expected to be invited to take my Brigade to help sweep up the mess after the armoured battles were over. But following the 100 hours ground campaign and the re-occupation of Kuwait it all went quiet as far as involvement by 3 Commando Brigade went. I was aware, too, of the plight of the Kurds, although now freely admit that I understood very little about the details of the problem.

The Crisis Unfolds

Then right out of the blue I was in amongst it, up to my armpits. Recalled one day from a NATO conference in Oslo by a friendly tip-off from a well-placed mate in the MoD I checked in with him on the phone while waiting for my baggage at Heathrow. 'It's a good job you called' he said. 'Tomorrow you have to be at Brize Norton by 09.00 to leave for Turkey with your recce party'. I looked at my watch. I had 15 hours to get to Plymouth, be briefed up for whatever lay ahead, pack my kit for an indeterminate period, say good-bye to my wife, join up with the rest of the recce group and get to Brize Norton near Swindon. I didn't spend much of that night sleeping.

How Did I Feel as I Set Off for Turkey and the Kurdish Problem?

- Very excited at the prospect of participating in such a crucial, high profile, multi-national operation.
- Very pleased that the Commando Brigade had been chosen to represent UK.
- Very confident in my team; both my immediate staff in my Headquarters and my units in the Brigade. This was purely a result of hard training and many shared experiences.

But all too Aware that:

- I and my subordinate commanders were all on a very steep learning curve. None of us had any previous experience in this part of the world.
- The eyes of the world's press were on us.
- A high cost would be paid for any sort of failure.
- Slightly apprehensive on a personal basis, but not overwhelmingly so.

Luckily for me the recce team had been well chosen. The team leader was an extremely able and experienced 2-star officer (he was later Chief of the Air Staff) who had held a key role in the Joint Headquarters at High Wycombe throughout the Gulf War. He was also wise, calm, enthusiastic and wholly supportive. My job was to try to identify a Concept of Operations – how should my Brigade be employed? And following on from that, what exactly should it consist of?

Within hours we were at Incirlik, a large NATO air base outside Adana in southern Turkey. The Coalition was slowly taking shape under the watchful eye of the Commander of the Combined Task Force (CTF), Lieutenant General John Shallikashvilli, US Army. Shalli too, was wise and calm; I saw him under pressure but I never saw him flustered. Like everyone else he had only just arrived, and like everyone else he was desperately trying to find out what was happening; in particular:

- Where were the refugees exactly?
- How many refugees were there in each 'camp'? Rumours suggested we were dealing with well over 1 million in total, maybe twice as many.
- What were the refugees most pressing needs? Food, water, first aid, medical supplies?
- What did the Kurds themselves think? How were they likely to respond to our efforts?
- What forces were available to the CTF? What were their capabilities? When would they arrive? Were they prepared to do what they were told, or were there some national restrictions on what they were prepared to do?
- What were the Iraqis doing? How were they likely to react to any of our operations?
- How would the Turkish authorities respond to a NATO-heavy coalition determined to 'save the Kurds'?
- How should we co-ordinate our efforts with the numerous Non-Governmental Organisations (NGOs) which were also flooding in to the area?
- What about the UN and its various agencies?

There were doubtless countless other considerations too. In short this was a highly charged and very complex operation, for which there were no contingency plans, no blueprint, and no neat solution. Meanwhile hundreds of Kurdish refugees were dying on a daily basis as the Turkish authorities held them on the Turkish-Iraqi border, high up in the mountains, miles from any sort of infrastructure, and the world's media were at least as well informed as we were, probably better.

For 5 days we tried as best we could to answer the questions above and come up with a sensible yet flexible plan which would mesh in with other members of the Coalition, principally the Americans. I then flew home to brief the Commander-in-Chief at High Wycombe, leaving most of my team in theatre. While I was in the air I was told that the Government had ordered my Brigade to start flying to Turkey – to respond to a plan which I was at the very moment still making and which had received no formal approval! For a while I felt as if I was a victim of some weird conspiracy designed to make an already challenging situation totally impossible!

At High Wycombe I presented the C-in-C and over 100 of his staff with my Concept of Operations. In essence it was approved. Then I returned to Plymouth to brief my people there and 2 or 3 days later returned to Turkey. By that time the CTF had been split into 2 Joint Task Forces (JTF):

- JTF-A was essentially made up of Special Forces, and was used to penetrate the refugee camps high up on the Iraqi/Turkish border to gather intelligence and to stabilise the situation as best as possible.
- JTF-B, under the command of a splendid US Army Major General (who by good fortune had started his military life as an enlisted Marine), formed the bulk of the Coalition forces, and was to generate a safe environment for the Kurds and to facilitate their well-being.
- I detached 40 Commando and a few others to JTF-A and with the majority of my Brigade immediately joined JTF-B.

My Command Role

My immediate task was to find out what was happening, and the imperative was very strongly to respond to it successfully. As in Belfast 5 years earlier I knew that my principal responsibility was to provide the leadership that my own people needed and expected. I also knew that I had to establish and maintain a sound working relationship with my new boss and his principal staff officers. As the HQ of JTF-B was itself being created on the hoof while individuals and units arrived often un-announced from countries various, they themselves had an extremely unenviable task and initially found it difficult to organise themselves let alone their units.

I spent my first few hours with the advanced elements of my Brigade camped in a huge field just on the Turkish side of the border. The following morning I pinched a landrover to drive across the border into Iraq to set up my small Tactical HQ close to the Commander of JTF-B. He was occupying the wreck of an Iraqi barracks on the edge of Zakhu, a city that should have had a population of about 80,000, but like all the other cities, towns and villages in North Iraq was virtually deserted. The whole place was a wreck – it had, after all, been a war zone for many years.

Developing the Plan

Over the next few days we concentrated on intelligence gathering and on beginning to build a huge tented camp on the plain outside Zakhu to house the refugees when they came down from the mountains. However, as we began to understand what was happening around us we came to realise that, if you happen to be a refugee, the last place you want to go is into a camp. Refugee camps are soulless places, and those that occupy them rapidly develop the status of being non-people. The Kurds knew this already, for most of them had previous experience of being de-personalised by Saddam's army, and thus far there was little to suggest to them that we would treat them differently.

So we not only had to gather intelligence but we also had to develop a plan to gain the confidence of the indigenous population. This was another real challenge for there was no apparent structure to the Kurdish community that was concentrating on trying to survive in a number of far-flung impromptu mountain 'camps'. I visited one early on and was amazed by the mass of people there, by the huge number of children, by the apparent lack of men, by the stench of human ordure and by the emptiness of the peoples' eyes.

Although in many ways the situation was very confused, we soon came up with our new outline Concept of Operations, which was:

- To provide immediate life-saving support to the hundreds of thousands of Kurds in the 'camps' on the Turkish border.
- To move the refugees down from the mountains – to the specific area from which they fled.
- To help the Kurds to re-settle themselves.

But how were we going to deal with the Kurdish 'command', if indeed such a thing even existed? By good fortune a Troop Commander in 45 Commando who was patrolling through the mountains with his men heard tell of 'General Ali' – clearly a Kurd of some substance. Over several days this energetic young officer tracked down Ali and persuaded him to come to Zakhu to meet General Garner, the Commander of JTF-B. The meeting was successful and enabled us to encourage Ali and his team to spread the word through the mountains that the Iraqi Army had gone and that the area was safe for their people. Meanwhile, rather than simply building a huge tented camp for these unfortunate people the onus was now clearly on us to carve out what became known as a 'Safe Haven' for the Kurds in their own country.

At very short notice and with alarmingly small quantities of ammunition (our heavy lift logistic supplies still had not arrived) we then advanced about 80 miles into Iraq. My Brigade, which was the largest, had the lion's share of the real estate, and I found myself responsible for about 5,000 square kilometres of mountainous terrain, with very few roads and absolutely no form of recognisable infrastructure. As we moved east we were initially confronted by groups of Iraqi soldiers, but thanks to the efforts of a small group of trained American negotiators there was very little Iraqi armed response. In fact they looked a sad, dejected and motley crew, but even so there were no guarantees initially that they would move aside for us.

Command Functions

Meanwhile I had to attend to all the normal command functions. In essence these all boil down to decision-making and effective communication up, down and across the chain of command. To achieve this I used my own Brigade staff as fully as possible. In simple terms this meant me giving my Chief of Staff (we had served together several times before and knew each other very well) broad guidance and then leaving him, with his team, to develop the detail. I trusted him totally and I think he trusted me. We met at least once and usually twice daily. While he ran the headquarters with consummate efficiency and style I foraged around on the ground talking to people – principally my own commanders, but also our other Coalition partners (Americans, Dutch, French, Spanish, Italian, Canadians, Australians and a few others) and, of course, Kurds. All this was to try to improve my understanding of what was actually happening, to inform me and others about the effectiveness of the decisions we had already made and of the relevant considerations for the next tranche of decisions. As in Belfast I found that, by spending many hours each day out and about on my patch I quite rapidly accumulated a comparative wealth of information and knowledge which was invaluable to my command function. Unlike Belfast though, I found that the longer I was there, and the more I learned, the more I realised how little I really understood about that troubled part of the world, its people, its history, its culture, its standards, its previous experiences and its expectations.

For the first few weeks it was incredibly hard work. I always tried to have 6 hours on my camp-bed at night and usually succeeded, for I knew how important it was to pace myself. I failed at least once and gave myself a nasty shock very late one night when I heard myself being totally unreasonable to an equally tired and probably more hard-pressed subordinate. Stress affects us all and it is vital that commanders recognise the symptoms in themselves, as well as in others. It was quite impossible to do everything that I felt I ought to be doing, so I had to force myself to plan and prioritise my own activities very carefully. I did this, as everything else, in conjunction with my excellent Chief of Staff. As I used him to put the flesh on the bones of all our planning and decision-making, so he depended on me to provide the bones of sufficient quality and quantity to enable him and his multi-disciplined team to be effective in providing the detailed instructions for the Brigade. We had over 4,000 people working for us, in an extremely, challenging part of the world and in a fast-moving, dangerous, real-life, high profile scenario which was attracting the close attention of the world's press and numerous leaders of governments. It was real pressure-cooker stuff, and needed careful thought and highly professional and very precise action in a very imprecise situation by a wide range of military fighting men, logisticians, engineers, aircrew, communicators and many others.

'Friction' – Unexpected but Inevitable

I had numerous other complications to deal with, including:

- A complex of Presidential Palaces in my Brigade area, which belonged to Saddam Hussein, and which, although not occupied, were Out of Bounds to us and were guarded throughout our tenure by the Republican Guard. This was quite an inconvenience!
- The whole question of our own logistic support. I had over 4,000 people who all needed food, water, sanitation, a huge range of supplies for vehicles, helicopters, radios and so on to say nothing of medical assistance when they were sick or wounded. Most of our daily needs were supplied by a Royal Fleet Auxiliary ship in the Eastern Mediterranean, but as the sea was about 400 miles away this also provided problems.
- All our supply lines came through Turkey, yet the Turkish authorities were very uneasy about the whole Safe Haven operation. They suspected a secret agenda whereby we were using the crisis as cover to re-arm the Kurds. This was very far from the truth, but it resulted in huge bureaucratic difficulties over our lines of communication.
- My superior chain of command was, in my opinion, less than perfect. On a minute by minute basis I was responsible for my tactical operations to the Commander of JTF-B. However, I was also responsible in a national sense to a British two-star officer and his somewhat ad hoc staff, situated 400 miles away alongside but not under command of the Commander of the Combined Task Force (i.e. the whole military Coalition) in Adana, Turkey. Movement between Adana and my Brigade took several hours and members of the UK staff

there, were very reluctant to be away from their own base overnight. So we received fleeting visits which were by no means always satisfactory. In addition, instructions passed to me by my two superior Headquarters were sometimes in conflict with each other, and on one occasion I received very specific but diametrically opposed orders at exactly the same moment!

- Once the Kurds started to re-occupy the Safe Haven they were able, to some extent at least, to return to some sort of normal life. Not surprisingly, when they were able to start thinking about more than just survival on a daily basis, personal differences and vested interests began to surface and within days we were faced, as with any other community, with a need for the law to be imposed. There were a few murders, and no shortage of other angry disputes. But how should we deal with this problem? Other than the tribal infrastructure, which was patchy at best, there was no Iraqi or Kurdish authority, no police and in effect no law. And we certainly had no right to impose our own law. Like so much of this operation none of us had ever considered this facet of a humanitarian operation before.

- As our ideas bore fruit and the Kurds came to place their full trust in us, they not surprisingly saw the glimmers of an opportunity for us to save them from their longer-term problem – namely Saddam and his hated regime. But our mission was, quite rightly, limited; we were simply there to get them down from the mountains and to enable them to get their lives together again as best they could. However, some of us were seduced by the romantic notion of 'Saving the Kurds' too, and many of my men felt somewhat guilty at the prospect of abandoning these people who still needed so much help. So I made a point of speaking to as many of my people as possible to try to put our efforts and the whole situation in some sort of context for them. I hope it helped them.

Success

Gradually we saw that our plan was working. First the leaders came out of the hills to see for themselves that the Safe Haven really was safe, then they brought the men-folk down, and then the families started trickling through. There were many thousands of them and they all wanted to go home, or as near to home as they could get, not to some ghastly tented camp. We set up 'Way Stations' for them, temporary bases where they could rest, feed, get some medical help, gain in strength and then move on. We had set a target of June 1st to have all the mountain camps vacated, and we made it with hours to spare. It was a complicated picture with many associated personal and collective crises and tragedies, but slowly the whole thing gained a momentum of its own and it dawned on us that our mission was achieving success.

Hansard records that Mr Tom King, then Secretary of State for Defence, made the following statement in the House of Commons in early June:

> There is no question but that we asked our Commando Brigade to undertake an incredibly difficult task. The brilliant way in which it discharged its duties is a measure of its confidence, training and absolute commitment. Others have played their part but none was better than 3 Commando Brigade. Its efforts and those of others undoubtedly saved the lives of hundreds of thousands.

By the end of June the situation was, to a considerable extent, stabilised. We, the military, had certainly done our job and it was time to hand over to more appropriate authorities such as UNHCR, the NGOs, and indeed the Kurds themselves. Over the last few weeks we thinned out our people and gradually made way for those who would be staying on. In effect, therefore, the operation just petered out, although a small reaction force was left across the border in Turkey for a while and Coalition aircraft have been patrolling the area ever since. I flew home in July, about three months after it had all started.

Overall Operation Haven was in most respects a great success. For some reason, though, it barely gets a mention in the British annals of post-Cold War military humanitarian operations, although the Americans and others recognise that it was a very significant event, as it was the first large-scale multi-national use of military force in response to a huge refugee problem. From my own perspective our success was due to 3 key factors:

- Training.
- A burning desire to succeed.
- Leadership at all levels.

How Did I Feel at the End of the North Iraq Tour?

- Tired, but very grateful for the opportunity to have contributed to such an event and satisfied that we had done a good job.
- Very proud that my young men and women had acquitted themselves with such distinction on such an important task, and that, quite rightly, they had been given the lion's share of the work in the field.

- Delighted with the way in which the chain of command had worked within the Brigade.
- Hugely impressed with the resilience and effectiveness of my Brigade staff, particularly their ability to respond so willingly and imaginatively to my broad direction.
- Pleased that, although my Royal Marines and the others in my Brigade had received no previous training in humanitarian operations and had no previous knowledge of North Iraq or the Kurdish problem, they had been imaginative, inventive and responsive to the totally unexpected scenario they had found themselves in.
- Convinced that much of our success was due to our relentless training regime, including the hard 2-month Arctic Warfare training period we had been through earlier in the year.
- Convinced, yet again, that in a real crisis commanders have to force themselves to concentrate on commanding and to use their time carefully and in a very focussed way.
- Relieved that I had learned about 'Mission Command' at the Higher Command & Staff Course, immediately before taking up my job as Brigade Commander. This new Army course had been designed to teach senior officers (for the first time, in the British Armed Forces) about field command techniques at brigade level and above. 'Mission Command' requires commanders to focus their attention on the Main Effort, and to communicate clearly to their subordinates what they are required to do without becoming involved in telling them how to do it.
- Sad that, from my point of view at least, the national chain of command had not worked as well as the Coalition chain of command. My perspective was that, unlike the Americans who delegated superbly and never interfered, my UK masters were not running along 'Mission Command' lines. I regretted this. In the absence of any significant military doctrine to deal either with general principles or operational specifics of the sort of operation that we were involved in, and with the consequent over-reliance on instinct, it probably was not surprising that relations sometimes became strained. I am pleased to say that the situation regarding doctrine is now much improved, for it not only exists but it is widely taught, understood and practised. Hence commanders are freed up to use their instinct within the doctrinal framework, which must be a better way.

Summary

This book is about decision-making in fast-moving, life-threatening crisis situations. In my opinion decision-making is *the* ultimate command function and activity. Just about everything I did in Belfast and North Iraq was specifically designed to enable me to make quality decisions. All the time I spent moving around my areas, talking to people, sharing their difficulties, experiencing their experiences, getting to know my boss, assessing what was and was not going on, trying to work out what it all meant; all of this was simply to enable me to fulfil my key leadership and command function of making decisions. In simple terms this meant trying very hard to make the best, or maybe sometimes the least worst, decisions concerning what I wanted my subordinates to do to meet the needs of the situation which prevailed. I felt I was best able to make such decisions if I was as well informed as possible about the situation and about them, my people. Having made my decisions, it was then my clear responsibility to communicate them clearly to my subordinates. Throughout the process I used my assistants, or subordinates, as much as I could, and delegated to them as much as possible. My job was to make the decisions (with my subordinates' help when appropriate); their job was to turn my decisions into actions.

However good your people are you are likely to be a more effective commander if you have jumped through as realistic a command hoop as possible as many times as possible before the real event. Such training exercises should be hard and should cause you to make mistakes. Unless you are put under real pressure in training you will not learn about the real demands of real incidents. Crisis is almost bound to be chaotic, and you must learn to survive and even thrive in chaotic situations. You must be able to 'click in' to crisis and chaos with your team. Of course there is no substitute for real experience, but that is no excuse for not training as frequently and as realistically as possible within the demands of your normal, routine life, whatever they may be. Luckily, and for a host of very good reasons, the Armed Forces dedicate a huge amount of time, energy and other valuable resources to realistic training and very thorough preparation. It works, as I discovered for myself in Ireland and with the Kurds.

A Rule-of-Thumb for Incident Commanders

1. *Prepare* yourself and your team as thoroughly and realistically as you can – before the crisis breaks.
2. Work hard to acquire an *intimate knowledge* of your key team. This will help to build trust and confidence both ways, thus enabling seamless delegation. You don't have to like your immediate subordinates – but it probably helps if you do.
3. Be very clear at all times about your *Aim* – whether it is imposed on you or you decide it yourself. This will focus all of your efforts.
4. Remember that your key personal activities are *decision-making*, and *communicating* those decisions clearly and concisely to those that need to know them.
5. When *directing* your subordinates, tell them What to do, not How to do it.
6. *Get about* amongst your people and your area of responsibility as much as you can – smell the air, share the danger, know what it is like out on the ground.
7. Give as much *notice* as possible of your intentions to your subordinates.
8. Establish and maintain effective *communication* with your higher headquarters – in other words respect the needs of your boss.
9. Remain *flexible* at all times, continually reviewing your Aim, your key decisions and your tasking.
10. Remember that crisis is likely to be *chaotic*, and there may not always be a 'right' answer. Recognise that Incident Commanders *must* be prepared to take risks and change their minds as the scenario unfolds.

Note

1. This whole sad saga is sometimes referred to by the Kurds as The Great Betrayal.

References

Bulloch, J. & Morris, H. (1992) *No Friends but the Mountains – The Tragic History of the Kurds.* London: Penguin.
McDowell, D. 91996) *A Modern History of the Kurds.* London: Collins.
Shaara, M. (1974) *The Killer Angels.* New York: Ballantine Books.

9 Airline Captain – The View from the Flightdeck

Mike Lodge
Boeing 747 Captain, British Airways (retd.)

Mike Lodge graduated from University College of North Wales, Bangor, and Tübingen University, Germany, before beginning his professional flying career at the College of Air Training, Hamble. He joined British Airways (then British European Airways) in 1966 as a Second Officer on Vickers Vanguards, and converted to Boeing 707 in 1972. He became a Training First Officer in 1975, gained first Command on B737 in 1980 and converted to B747 in 1986. Subsequently he became a B747 Line Trainer, then a Base Trainer and retired as a Standards Captain in 1996. He now works in aviation safety, on European research and consultancy projects.

A Pilot's Lot

There's an old flying adage that describes a commercial pilot's working life as 95% boredom and 5% panic. Looking back over my four decades of involvement in aviation, I suspect that comes pretty close to the truth for me and many of my colleagues.

What we learn from those moments of panic is probably the key to our continuing survival in an environment that can be at times extremely hostile. Remember, this is a workplace that may, at one minute, have us perched serenely many miles above the glories of Greenland and then see us battling with a tropical storm on the approach into Miami, unravelling the complexities of a rush-hour arrival at Chicago O'Hare or sliding onto the icy ramp at Anchorage, Alaska. Oh for more boredom!

Every pilot will have his own stories to tell and just like the old sailor holding court in the corner of his favourite bar, he will probably relish – and embellish – the telling. And like the old salt, he will have had experiences that he would rather keep to himself. Maybe they were too embarrassing to relate or maybe the fear of stirring up best-forgotten emotions kept the lips sealed. Yet only by sharing our experiences can we speed up the pace of learning and thus increase our chances of surviving into retirement. It is in that spirit of exchange I begin this journey through my flying career. There is absolutely no intention of raking up scare stories or of sullying the reputations of characters and organisations I have worked with over the last forty years. It has been my privilege to fly alongside some of the best pilots, flight-engineers and cabin crew in the business and I hope that passing on some of the lessons I have learned from them will go some way to repaying the debt I owe to their skills and expertise. If the best can justly serve as icons to aspiring aviators, then let us not forget the other lesser mortals, who too have a role in teaching us perhaps how not to be. I hope to show both sides of this important learning equation.

Luck be my Lady Tonight

A famous golfer, Arnold Palmer, I think, was once chided by a spectator for being 'lucky' after he had just holed an unlikely 30-foot putt. The golfer replied how strange it was that 'the more he practised, the luckier he became'. There is no doubt that luck plays a crucial part in our lives. To what extent we 'make our own luck' is debatable and much philosophical argument has focused on the influence we can exert over the way our paths turn from one course to the next. I am convinced that luck has played an important part in my career. I have no religious beliefs to which to ascribe my good fortune at critical moments in my life. Some, like the golfer, I could put down to 'practice' and the occasional sound judgement but others defy logical interpretation. Maybe one day all will be revealed.

We all have events in our lives that we later realise were pivotal in determining the way things were to turn out. It seems to me now with the considerable advantage of forty years of hindsight that many apparently insignificant events form the basis for those judgement processes that we take as almost automatic in later life. As a senior airline Captain commanding an aircraft costing many millions of dollars, with the lives of four hundred or more trusting human beings in your hands, it is very easy to forget that the way you react to situations and cope with challenges has often been determined by things that happened many years before. Our early years of apprenticeship seem to me to be crucial to how we finally turn out. Perhaps a short look back into my own beginnings might shed some light on this.

A Fledgling Flies

My first attempted entry into the world of piloting was stillborn in the late Fifties, due largely to an amazingly misguided statement made by the then Minister of Aviation, one Duncan Sandys, that the future lay only in missile technology and the manned aircraft was soon destined to become a museum-piece. On the advice of my Headmaster I abandoned thoughts of getting into the Royal Air Force and went on to University to study Modern Languages. Here luck played its opening hand because my degree course involved spending eighteen months in Germany and whilst there I joined Akaflieg Tübingen – the Tübingen University gliding club. This was one of the best decisions I ever made. I got airborne for the first time in a glider and tasted the heady delights of flying. I soon realised that this was still what I wanted to do as a career but I had no idea how I should go about it. Before returning home to complete my degree I sent off an application to the College of Air Training, Hamble, where cadets were taught to become pilots for BOAC and BEA. I was rejected on the grounds I was 'too old' (I was then 23!). To my great surprise a few months later I received a letter suggesting I should re-apply to the College, as they were now considering a course purely for University graduates of slightly more mature years. After a series of aptitude tests and interviews I found myself at Hamble, along with eleven fellow graduates on a specially tailored experimental course. I felt very much the outsider in those early days. All my colleagues except one were science graduates, some had aeronautical degrees and nearly all had flown with University Air Squadrons. It took me some time to find my feet in the lessons on thermodynamics and electronics – all very new to a German scholar – but again luck stepped in on my side. Since my early teens I had nurtured a fascination with old cars, particularly Austin Sevens. I had learned to drive on a pre-war Morris 12 and had taken apart and rebuilt several Austins before arriving at the College in my 1931 Austin Swallow. This experience put me ahead of the pack when we started learning about how aircraft engines actually worked – nothing much had changed since before the war as far as the pilot examinations were concerned. I realised that, although I would need to work harder than the rest, mine was not a lost cause. When we put on our flying suits I soon noticed that my progress was as good as theirs and because I had no preconceived notions about how things should be done, I was often better placed to learn the new skills. It was a time of great pressure – the failure rate in those days was high and the standards required to pass the final flying test seemed, to us at least, to be dependent on the prevailing commercial needs of the airlines. But pass I did – in fact I was chosen as the top cadet of my course.

There were many lessons to be digested during those twelve hectic months but two particular events stand out that I can still have nightmares about!

First some background. Our graduate course at Hamble was carefully structured to provide the maximum amount of training value in shortest possible time. We alternated one day of ground school with a day of flying and we therefore made rapid progress but the price that had to be paid for this intensive regime was continuous assessment by our instructors. No freewheeling was allowed and a serious flying mistake could mean the end of a dream. In this atmosphere of extreme pressure to succeed there was certainly no incentive to speak up and admit to errors of judgement or aircraft handling – if they had not been witnessed by an instructor. Close friends in the sanctuary of the local pub might hear of a slightly 'hairy' event or two but many of them were never mentioned until after the longed for graduation was safely in the bag. So it was with me.

Trainee pilots spend their first tentative sorties in the attentive company of their instructors but when they are judged ready for their first solo flight and they finally break free, an exciting new world beckons. My debut was completely uneventful, unlike a course colleague's, that forced him to face the challenge of a badly under-performing engine with ignition problems. A death-defying low level circuit and a mercifully safe return to *terra firma* are surely imprinted forever on his memory. I can still see his Chipmunk disappearing behind the college buildings as he limped around the airfield, only to reappear what seemed like aeons later for a very creditable landing. With the solo hurdle safely negotiated there begins a period of consolidation and many hours of solo cross-country and circuit flying. Not just in the daylight of course and here I return to my first two 'character building' episodes.

I was never completely happy during the *ab initio* part of my Hamble training. The Chipmunk trainer was a worthy testimony to British aircraft design but it was cold and uncomfortable and unless you were of the breed that revelled in violent aerobatics or crossing the Solent inverted (I was not), the progression to the twin-engined Piper Apache was eagerly awaited. Flying suits were discarded for blazers and flannels and the world looked far more civilian – and civilised.

A spring evening near the end of the course found us in gaggles 'circuit bashing'. 'Solo night circuits' my logbook confirms. After four or five successful touch and go's, I climbed away on the sixth feeling rather smug and professional. 'After take-off checks', I recited to myself and began to complete the memory drill. To my great surprise (an understatement of considerable proportions) the left engine lost power and stopped. We had practised this eventuality many times before of course but always with the reassuring presence of an instructor and secure in the knowledge that it was not for real. This time it was different. I carried out the engine failure drill and continued straight ahead out of the circuit while I planned what to do next. The most pressing consideration seemed to be what I would say to the Air Traffic Controller as I turned into the circuit to land. Irrationally I was not concerned about the one-engine approach and landing or the distinct possibility that the other engine might also join its now silent partner and 'cease to be'. No, the only thought was to get the R/T call right. I mentally rehearsed it a couple of times:

Hamble Tower: X-ray Charlie downwind with port engine failed for full stop.

I took a deep breath and nervously began the call. A kindly angel standing at my shoulder made me look down at the magneto switches for the left engine and they were OFF. I must have switched them off by mistake when I did the After Take Off checks minutes before. I quickly tried to restart the engine and, Lycoming be praised, it roared into life. Instinct told me to continue as if nothing had happened, although I could not believe no one had noticed me weaving about after the engine 'failure' and heading to the north with one at full power. But there it was. Nobody had seen or heard anything unusual and so I vowed to keep it to myself; a lesson learned. I never mentioned it until several years later when flying stories were being swapped at a party. But exactly what was the lesson I had learned? A cynic might say 'don't admit to anything they can't prove' but the young aspiring pilot might be better advised to take away with him the need to prioritise his actions in critical situations. Nearly thirty years afterwards I was painfully reminded of that simple rule when flying a 747 over the Northern Territories of Canada but we will come to that later.

Now a short anecdote, which I am sure pilots will find familiar. 'Getting disorientated', 'losing the picture', 'losing your place', 'losing situation awareness' – we have all heard these phrases before in many different contexts. Once, in a glider without blind flying instruments, I had fallen out of a cumulo nimbus cloud in a tightening spiral dive. I should not have been in that cloud of course, but my first real encounter with this potentially lethal phenomenon was on a beautiful clear starry night over the Isle of Wight. I was returning from a night cross-country exercise, again mercifully solo. I was very pleased with myself at having completed the task without any of the traditional cock-ups and was looking forward to my bed. The island was spread out beneath me as I approached the Hamble Estuary, the lights of Southampton and Cowes sparkling in the cold night air. Suddenly a red light loomed up at my height – dead ahead; it seemed only a few yards away! I banked violently left and attempted to dive away underneath the traffic. I lost a lot of altitude in the manoeuvre and took some time to recover the aircraft back into stable flight. I looked around to see what the devil I had only just avoided. But the sky was empty. Down on the ground some miles away an innocent refinery chimney burned off surplus gases. I had just had an airmiss with the Fawley Flame! I was not the first to suffer this optical embarrassment and surely not the last. Crews have lost their lives since flying began due to this most human of mistakes: *loss of situation awareness*. I was 'lucky' again. Over the years many were not.

Now in 2001, many of us involved in pilot training are working hard to increase our understanding of this potential killer and to offer ways of loosening its insidious grip on the unwary (see Chappell, 1997; Endsley & Garland, 2000). As I write, a major new initiative will shortly be in place at British Airways to heighten pilot situation awareness, particularly amongst those crews flying ultra-modern 'glass cockpit' aircraft with their computer-driven displays and 'fly by wire' technology. There are also tentative moves to define the ideal profile of a *situationally aware* pilot before he or she embarks on a career in aviation but these selection criteria are highly controversial.

I have tried hard to think back over those distant days at College and to remember any formal training in the art (or science?) of decision making and coping in crises. Memory is a notoriously unreliable witness but I am fairly certain that such subjects were probably wrapped up in military jargon like 'airmanship' or 'leadership' or 'command potential'. All things, you, a would-be 'officer' were either born with or not. They could not be taught. Bear in mind the background of the instructors in the Sixties – both ground and flying – was almost exclusively ex-military and it is not surprising that their approach mirrored the way they had been trained at the end of the Second World War. (On our first day at Hamble we were greeted by the Principal – a rather fearsome antipodean ex-Air Vice-Marshal – who exhorted us all to concentrate on our flying and not the ladies of the village, with the timeless phrase: 'You can't f... *and* fly!' My anonymous colleague's response that 'sparrows do quite well' was, fortunately for him, not heard by that refined gentleman.)

Whilst on the subject of training techniques in those early days, I should point out in defence of the Hamble 'machine' that many hundreds of excellent pilots were produced and that the majority of the senior pilots at British Airways are evidence of the product. Recruitment criteria and the way embryo pilots are developed into the present day aviators have of course moved with the times. And yet the legacy of the old selection methods and the training given at the College of Air Training and its successors remains with us. The guy who set off fire extinguishers in his victims' bedrooms or threw unfortunates into the pool at midnight may well be wielding his Route Check Captain's pen over a pilot's career as you read this....but I digress.

With our flying tests and exams behind us we waited for our fate to be decided. Would we go to BOAC and its longhaul fleet of VC10's and B707's or would it be BEA and shorthaul Tridents and Vanguards? – the latter not being the preferred option for most of my course mates. The way our careers would develop was ultimately decided by a highly sophisticated and peculiarly British selection method: eleven names in a wastepaper bin and the first five out to go to BOAC! With my recent experience of living in Germany and my degree in the language, I hoped for BEA and many happy night stops in Berlin and Frankfurt. The first wish came true and BEA it was to be. The second however remains unfulfilled to this day. Even though I started out in the European division before moving to longhaul I was never on the fleet that serviced the German routes. Probably to the benefit of my marriage if nothing else...

'Benign Deities?'

There is no doubt in my mind that the first four or five years of airline experience will determine how a pilot is likely to progress through his flying career. Those who struggle to cope during that early stage will still be struggling thirty years later as retirement looms. This is why I believe it is essential for the role models, the Captains and senior co-pilots, to set the right example to their junior crewmembers. How was it ɪor me?

It is sometimes difficult now to appreciate just how hierarchical commercial aviation was thirty years ago. The captain saw himself very much as benign deity. (Some were more benign than others.) It would be very unusual for him to explain his decisions to his crew and it would take a very brave second officer to question a judgement or action. Not being party to the process made learning a hit and miss affair. If you were lucky enough to fly with an enlightened commander who confided in his co-pilots then there was a chance of picking up some valuable pointers. Let's look at two incidents where I was involved as a co-pilot and see what I mean.

The Boeing 707 was a wonderfully rewarding aircraft to operate and I am eternally grateful that I had such a long and intimate relationship with it. But it had some characteristics that could lead you into trouble. One such was the hydraulic system. It was a relatively highly stressed jigsaw of pumps and pipe-work that was not really designed for the demands of the shorter routes we were asking it to cope with. Hydraulic failures were fairly common and though not theoretically life-threatening they required thought and planning to ensure a safe outcome. We were approaching Heathrow early one Saturday morning on a flight from Rome that had originated in the Middle East. Just after we had lowered the undercarriage, one of hydraulic pumps decided to fail. The correct procedure involved a fairly lengthy checklist session culminating in manual winding of the nosegear locking system. This would all take time and mean a trip into the holding pattern to the south of Heathrow. The captain decided unilaterally that he wanted to continue straight in to land. No emergency was declared and we landed safely a few minutes later without completing the full checklist. So far so good but apart from the risk of the nose gear collapsing before we could fit the ground locks, we also had reduced brake capacity after clearing the runway. I climbed down through the underfloor electric's bay, out onto the taxiway and fitted the locks. (The undercarriage locks are large forged pins that, once inserted in the correct part of the mechanism prevent its collapse when there is no hydraulic pressure.) The tractor was duly called to tow us to the ramp and it finally arrived. In the semi-darkness I turned to climb up into the flightdeck and was promptly struck in the back by the large wing mirror on the tractor. No brightly coloured tabards in those days. It would have been a very unheroic death – flattened by a tractor!

As an example of how to make decisions and manage a minor crisis this incident has nothing to commend it. Even the most basic adherence to the Standard Operating Procedures was missing. On the positive side, the passengers were delivered safely to the terminal with a minimal delay, blissfully unaware of the dramas up front.

But now we'll review another B707 event. One of the Third Pilot's (Flight Engineer's) many duties immediately after getting airborne was to monitor the brake system indications to check whether a brake unit could have been binding during the take-off roll, giving the risk of an undercarriage fire. This was a well-established routine and I did the check as usual on the departure from Heathrow to Malta. Unusually, the indicator showed a 'premature release' on one of the wheels, meaning either a faulty warning or a real brake fault. The captain on this occasion was a very different animal. He had recently converted onto the B707 but had already been a training captain on another aircraft. He suggested we treat this as an exercise in emergency planning. He immediately involved me and the other co-pilot in formulating a plan of action. We contacted our engineering experts by radio and they advised it was probably a false warning and that we should continue to our destination where the system would be checked on the ground. The captain agreed to proceed to Malta but, just in case the warning was an indication of a real problem, he decided we should carefully consider all the possible ramifications if we were to land at Malta with seized brakes. He called up the Chief Stewardess and briefed her on the situation. She relayed the plan to her crew. It was decided not to say anything to the passengers at this stage, as there was little anyway that they could do until after landing.

The flight continued uneventfully to Malta where the captain told the Air Traffic Controllers of our potential problem. At that time the island was a highly active military base and contained the core of our Mediterranean Quick Response capability, so we felt confident they could handle any emergency that might arise. We landed safely with no apparent difficulty and began to return slowly along the parallel taxiway towards the terminal buildings on the far side of the airfield. As we passed the entrance to the RAF area we noticed it was getting progressively more difficult to taxi the aircraft. The brakes were beginning to seize and overheat. ATC warned us that under no circumstances would we be allowed to block access to the military ramp and so we were forced to continue. As we arrived on our parking position the brakes finally reached melting point and the hydraulic systems around the undercarriage legs burst and caught fire. Thanks to the careful planning and briefing by the captain, the RAF fire crews were already in position with their powder extinguishers and silver suits. The passengers evacuated the aircraft in clouds of smoke and dust and we then started filling in the inevitable paperwork. The procedure we had followed later became known to us as 'CRM' – Crew Resource Management: *'using all the available resources – information, equipment and people – to achieve safe and efficient flight operations.'*(Lauber, 1984,p. 20) (See also Salas et al., 2001; Wiener et al, 1993; Crichton & Flin, this volume).

The fact that I still remember these two incidents so vividly is testimony to their influence on my development. I hope that I was able to live up to the example set in the second. I am sure that I never deliberately rode roughshod over any of my crewmember's safety concerns and I have always tried to train young pilots to respect the gut feelings and reservations of their colleagues in critical situations. The final buck inevitably stops with the Captain but the route it takes to get there may well fatally influence the outcome of the events. Teamwork will usually assist us in finding an optimal course of action. Command can surely be lonely but we should not seek a forced isolation.

When we consider decision making in a piloting context, we are often talking about reacting quickly to a given set of circumstances in a tactical way. We often do not get the opportunity to plan strategically. Of course those tactical decisions may have a significant influence on the outcome. Consider this sequence of events that took place on St. Patrick's Day, 1977 at Prestwick, in Scotland.

We were starting the last day of a weeklong Base training detail, converting pilots from other types onto the B707. The week had gone well; our two trainees – one young First Officer and one senior Captain – had performed very competently throughout and the Training Captain and myself, then a Training First Officer, were confidently looking forward to going back to London after lunch with two happy, newly qualified B707 pilots. The weather was not unusual for the time of year at Prestwick: broken cloud at 2000ft and a stiff southwesterly wind. The planned exercise was for the First Officer to practise a number of simulated engine failures at or shortly after V_1. (The speed at which the pilot must decide either to reject the take-off or continue into the air). I was sitting on the observer's seat, supervising the converting captain who was carrying out the Flight Engineer's duties. Shortly after lift-off, the No.1 thrust lever was pulled back by the Training Captain to simulate the engine failure. It is likely at this point that the wind speed had increased considerably above the figure we had been given a few minutes before and combination of the swing to port from the loss of thrust from the left outboard engine and the natural weather cocking of the aircraft into the wind from the right, caused momentary confusion in the First Officer's mind. The need for rapid rudder input was not immediately appreciated and the aircraft began a sharp roll to the left and sideslipped towards the runway. The Training Captain took control but was unable to prevent a violent collision with the ground. During this and the subsequent second impact, the engines broke away from the wings, the undercarriage collapsed, the aircraft's back was broken, the fuel tanks ruptured and the 40 tonnes of kerosene ignited. Not much time for a strategic plan here. But the fact that we all four survived the accident with only minor injuries was only partially down to our good friend 'luck. ' Clear thinking by the Captain in directing the escape through the starboard door away from the burning wing and the speed with which we jumped from the wreckage before the cabin fire reached the flight deck were also crucial. One broken leg for the trainee captain and a painful dowsing in kerosene for my genitals were the only physical injuries that day but there were many lessons learnt from that brush with an early cremation. The Prestwick crash signalled the end for traditional Base-training and gave a much-needed impetus to the comprehensive use of simulators for this very dangerous activity. The training of instructors was also vastly improved and the risks of practising engine failures in crosswinds were underlined. I took with me the realisation that things can change very rapidly at the sharp end. A moment's inattention at a critical phase of flight and your world can fall about you.

When I arrived home from a couple of days interrogation by the AIB Inspectors, my wife, who was then six months pregnant, told me she had discovered a breast lump. It turned out to be the start of the hideous disease that subsequently took her and the unborn child from me. How much can happen in one short week....?

A Cunning Plan...

It's time to move on to later in my career and to review a few incidents where time was not so critical and a properly crafted plan was at least theoretically possible.

Tyre failures on take-off are mercifully rare these days with modern high performance materials but they still occasionally happen and can be extremely dangerous, especially if misdiagnosed as a power plant failure, which they can often resemble.

Taking-off from Naples Capodochino in a B737, I felt a loud thump and shudder through the controls as nose lifted from the runway. It was immediately followed by a sharp swing to the right. 'Starboard Engine failure' I thought, but a quick glance at the engine instruments showed everything working normally. My heart rate began to reduce somewhat as we climbed safely away over the inevitable hospital complex near the airport. I recalled a similar sensation some years before leaving Madrid Barajas and deduced, correctly as it turned out, that we had experienced a tyre failure. I suggested to my co-pilot that we should leave the undercarriage extended to cool while we checked with ATC. It took about twenty minutes before the search confirmed the remnants of one tyre tread scattered near the runway. We were now well on the way northwards to London and there was little point in stopping at Rome or Milan. We assessed the possible damage to the aircraft and decided to continue as planned to our home base, London Gatwick. A proper cabin crew brief was now obviously essential in case of a possible directional control problem on landing and an emergency evacuation. Although at that time Gatwick had only one runway, ATC allowed us to plan on landing there. A low-level fly past with binoculars trained on our wheels showed the offending tyre to be still inflated although treadless. We circled round and landed safely. Later inspection showed damage to the starboard flaps caused by flying rubber debris – the source of the swing on take-off. Not a major incident but none the less satisfying for me as new Captain.

I referred near the beginning of my 'tale' to an early example of my getting the priorities wrong in an emergency. Three decades later I was reminded of that trap for the unwary.

We had departed the West Coast of the United States for Heathrow and were routed high into the Canadian Arctic in early spring. It was all very routine and we had reached the 'boring 95%' portion of our trip. My co-pilot began to suffer serious stomach cramps and he was soon looking very unwell. We suspected appendicitis. We decided to move him into the crew rest area aft of the flight deck where a bunk was fitted. The Flight Engineer helped him into the bunk with the aid of a stewardess and we called for a doctor over the PA. One fortunately appeared within a couple of minutes and began an examination. While this was going on I started to consider our options, if our co-pilot needed urgent hospital treatment. I looked over the charts of the area and searched for an airfield that was suitable for a heavy B747 to land and then take-off again with enough fuel to reach London. I spent some time completely engrossed in this until the Flight Engineer returned with the news that the First Officer was not dangerously ill and could wait until London for medical attention. I proudly told him I had found two airfields nearby where we could have dropped off the co-pilot and then continued straight on to Heathrow. Flight Engineers are not known for their reticence when faced with pilot stupidity and this one pointed out the rather obvious fact that without our First Officer, we were not going anywhere that night and I had rather wasted my time worrying about it. I wonder how many decisions have been made over the years with similarly shaky logic? 'Keep your eye on the ball' applies here as in many other activities.

A rather more complex situation confronted us once on a flight from Moscow to Tokyo some years later. As is often the case with technical failures on large aircraft, this emergency developed because we had been despatched – quite properly and within the 'rules' – with one known unserviceable piece of electrical equipment and several later seemingly random failures left us very exposed over an extremely inhospitable terrain. We had left Moscow with one of our four electric generators reportedly unreliable. As expected on the climb, the generator showed indications of overheating and my Flight Engineer carried out checklist items to remove its electrical load. I suggested we might as well disconnect it completely, as it was no longer any use to us but he prudently declined and left it idling. Some hours later over Siberia in deepest winter, the cabin crew reported that the lights in the rear cabin were flickering in a strange way. The Flight Engineer confirmed that he had already noticed an erratic surging on the three remaining generators. While he was investigating this problem, a second generator failed, followed quickly by a third. We were now reduced to one source of generated power and things began to get interesting. Life got difficult in the cabin because, apart from the lack of proper lighting and in-flight entertainment, the galley power had now been switched off to protect essential instrument and navigation supplies. My concern now moved swiftly to the possibility of a total generator failure and the prospect of finding a place to land in pre-Glasnost Russia, in winter, at night with perhaps thirty minutes of battery time left to us. We heard the welcome sound of a fellow B747 high above, on his way direct from London to Tokyo – a *very senior* Captain's route at that time – I asked if perhaps he could slow down a little and let us visually format on him until we reached the coast? He seemed oblivious to the urgency of my radio request and continued blithely on his senior way. The Flight Engineer's sensible suggestion of some hours before to keep the first generator in reserve proved our salvation. By returning it carefully to service, he was able to restore one of the other recalcitrant power sources and gradually life returned to normality. I have often wondered how we would have coped if things had taken a more fateful turn. I believe we covered the options as they revealed themselves but to stay ahead of the game in rapidly changing circumstances is the real challenge of aviation.

Today all our young pilots have the opportunity to build on the expertise and wisdom of their elders. They are trained from the outset to manage situations and not merely react to events. Decision making accessories are now *de rigeur:* BA has DODAR (Define problem, Options, Decide, Act, Review), Lufthansa has FOR-DEC (Facts, Options, Risks, Decisions, Execution, Check – Hörmann, 1995). Others have developed even more exotic acronyms but these should not be seen as the way to reach decisions but rather as a checklist after other more intuitive processes have been employed. (See also Larken, this volume and Klein, 2000 for similar views on prescriptive decision methods). Even error-management strategies are now properly trained, much to the chagrin of some old-school instructors, who still hold the conviction that we should teach pilots to be *infallible* (just like them).

There is now a wealth of academic research and operational experimentation all focused on optimising positive behaviours and crew interaction thus promoting enhanced safety (e.g. Helmreich, 2000; O'Connor et al, in press; Orasanu, 1997, 1999; Reason, 2000). The young pilots have got it made... the differences between their world and the one that confronted me forty years ago are staggering. And yet....

Summary

- Be situationally aware – stay ahead of the game in rapidly changing circumstances...
- The old CRM message: 'use all available resources..'
- Respect the gut feelings of your colleagues...

- Prioritise your actions in critical situations.

References

Chappell, S. (1997) Managing situation awareness on the flight deck: The next best thing to a crystal ball. www.crm-devel.org/resources/paper/chappell.htm.

Endsley, M. & Garland, D. (2000) (Eds.) *Situation Awareness: Analysis and Measurement.* Mahwah, NJ: Lawrence Erlbaum.

Helmreich, R. (2000) Managing threat and error: Data from line operations, In B. Hayward & S. Henderson (Eds.), *Proceedings of the Fifth Australian Aviation Psychology Symposium, Manly, November.* Sydney: Australian Aviation Psychology Association.

Hörmann, J. (1995) FOR-DEC: A prescriptive model for aeronautical decision making, In R. Fuller, N. Johnston & N. McDonald, (Eds.) *Human Factors in Aviation Operations.* Aldershot: Avebury.

Klein, G. (2000) Decision skills training for the aviation community, In B. Hayward & S. Henderson (Eds.) *Proceedings of the Fifth Australian Aviation Psychology Symposium, Manly, November.* Sydney: Australian Aviation Psychology Association.

Lauber, J. (1984) Resource management in the cockpit, *Air Line Pilot,* 53, 20-23.

O'Connor, P., Hörmann, J., Flin, R., Lodge, M., Goeters, K-M., and the JARTEL group (in press). Developing a method for assessing CRM skills: A European perspective. *International Journal of Aviation Psychology.*

Orasanu, J. (1997) Stress and naturalistic decision making, In R. Flin, E. Salas, M. Strub & L. Martin, (Eds.) *Decision Making Under Stress.* Aldershot: Ashgate.

Orasanu, J. (1999) Pilot error detection/ monitoring and challenging in the cockpit, In R. Jensen (Ed.) *Proceedings of the Tenth International Aviation Psychology Symposium, Columbus, April.* Columbus: Ohio State University.

Reason, J. (2000) Heroic compensations: The benign face of human error, In B. Hayward & S. Henderson (Eds.) *Proceedings of the Fifth Australian Aviation Psychology Symposium, Manly, November.* Sydney: Australian Aviation Psychology Association.

Salas, E., Bowers, C. & Edens, E. (2001) (Eds.) *Improving Teamwork in Organizations. Applications of Resource Management Training.* Mahwah, N.J.: Lawrence Erlbaum.

Wiener, E., Kanki, B. & Helmreich, R. (1993) (Eds.) *Cockpit Resource Management.* San Diego: Academic Press.

10 Prison Governor

Andrew Coyle
Kings College, University of London

Andrew Coyle is Director of the International Centre for Prison Studies in King's College London.[1] He previously had 25 years experience at a senior level in the prison services of the United Kingdom. While in the Scottish Prison Service he was Governor of Greenock, Peterhead and Shotts Prisons. Between 1991 and 1997 he was Governor of Brixton Prison in London.

In 1987 the Scottish Prison Service established national Incident Command Teams as a response to a series of major riots which were threatening the stability of the entire service. Andrew Coyle became commander of the first of these teams. This chapter is a personal reflection on his experiences in this role.

www.prisonstudies.org

The Scottish Prison Service in Context

Unlike the police and fire services, which are organised on a local basis, the Scottish Prison Service (SPS) is a national organisation. During the period we are concerned with, it was part of the Scottish Home and Health Department within the Scottish Office. The Director of the Service was a senior civil servant with an administrative rather than an operational background. His senior deputy was the Controller of Operations. Traditionally a former prison governor, who was responsible for all operational matters, held this post. The 21 prisons in Scotland were under the command of governors, all of them career prison people. They reported directly to the Controller of Operations and through him to the Director.

176

The smallest prisons held less than 100 prisoners. Barlinnie in Glasgow was the largest, holding 1000 or more. Other large prisons held between 500 and 700 prisoners (Scottish Prison Service, 1998–99).

In democratic societies people can be deprived of their liberty and locked up in prison only after a judicial process. There is also a legal framework defining the conditions under which they should be kept in prison. In Scotland in the late 1980s these were contained in the primary legislation, the Prison (Scotland) Act, which was very general in terms, and in the Prison (Scotland) Rules, which were slightly more specific. Administrative Standing Orders contained much more detail about how prisoners were to be treated and prisons were to be managed. All of this meant in organisational terms that policy matters were decided at central level but that governors were left largely on their own when it came to managing day to day activities in each prison.

By definition, prisons are coercive institutions in so far as the prisoners do not choose to be there. The primary task of prison staff is to ensure that prisoners complete the period of deprivation of liberty set by the court. This does not mean that prisons are invariably places of high tension. On the contrary, the reality of much of prison life is surprisingly relaxed. Although prisoners would prefer not to be there, almost all accept the inevitable and want to serve their sentences with as little fuss as possible. Similarly, prison staff will normally do all that they can to create a positive atmosphere within a prison. It is this relationship between front line staff and prisoners which is the key to a well-managed prison (Cressey, 1961). The tradition in the Scottish Prison Service, since it was established in its present form in 1878, has been one of stability and of an understanding between staff and prisoners. For a short and traumatic period in the late 1980s that stability and understanding came under real threat (Coyle, 1991).

A Changed Scenario

Until 1987 the options facing prisoners serving long sentences in Scotland were limited. Generally speaking, those serving their first sentence were held in Edinburgh Prison; those who had served previous sentences but were likely to conform to the necessary discipline were held in Perth Prison; and those who were thought unlikely to conform or who posed a particular threat of escape were held in Peterhead Prison. In general terms the system was very simple and predictable. This element of predictability is an important feature in prison life and the consequences of tampering with it can be serious.

By 1986 there was a severe imbalance in available accommodation in Scottish prisons and it was decided to embark on a major re-allocation of prisoners. In the course of a few months in 1987, under a project given the unfortunate title of 'Grand Design', 25% of all Scottish prisoners, the majority of them serving long sentences were transferred to another prison. In the short term, all went well and the transfers were effected without major disruption. In the longer term, it is hard to avoid the conclusion that the major incidents that took place in Scottish prisons between the autumn of 1987 and 1988 were influenced by the instability engendered by the disruption of so many enforced and unpopular transfers of long-term prisoners.

Under the previous arrangements any prisoner who was considered to be a serious threat to good order in any prison could expect to be sent to Peterhead Prison. This was an unsophisticated and at times unjust style of management but it was efficient. In general terms there was very little disruption in Scottish prisons with the exception of Peterhead. The much higher ratio of staff to prisoners at Peterhead ensured that even there, disruption was kept to a minimum and was likely to be short-lived. In this prison staff knew their place and prisoners were expected to know theirs. Most prison systems have a prison like this, one that is known as the place of last resort, the end of the road, the 'hate factory'. Prisoners who caused problems in other prisons knew that they would quickly be transferred to Peterhead. The price to be paid was a steady flow of unrest in Peterhead Prison. For example, between 1984 and 1985 there were three major incidents in Peterhead: a roof top siege, a hostage incident and a serious escape attempt (Coyle, 1994; Wozniak, 1989).

At the end of October 1986 there was a major incident which represented a sea change on two counts. In the first place, it happened in Edinburgh Prison, which had never previously faced major disruption. The incident was also different because of the scale of the violence involved. In the years immediately prior to this there had been an increasing number of hostage incidents in prisons in the United Kingdom. These were a relatively new phenomenon on the prison scene in this country and tactics were being developed on how to deal with them. Until the October incident in Edinburgh, all of these events had involved one or at most two prisoners taking a single member of staff or another prisoner hostage in a relatively confined space such as a prison cell. The techniques being developed for managing these incidents reflected these scenarios. In Edinburgh, for the first time, prisoners took over an entire living unit, in this case an accommodation hall with three floors and over 100 cells. The hall was badly vandalised with many of the internal sanitary and other fittings pulled out and thrown into the central well; the final repair bill was put at £300,000. Negotiating in a situation like that was a completely new experience for the staff involved as were preparations for intervention (*The Scotsman*, 29/10/86 and 30/10/86).

This was the first such incident but it was certainly not the last. Ten days later a similar and, if anything, more violent hostage incident happened in Peterhead Prison. In this one an officer was held on the roof of a main hall and was photographed by the world's media sitting in a terrified manner in the middle of a group of menacing prisoners. The incident came to an end when fire broke out and the hall was almost destroyed. The officer was released physically unharmed and all the prisoners were safely accounted for. The damage this time was £500,000. Three of the ringleaders subsequently received additional sentences of 10 years each (*The Glasgow Herald*, 13/11/86, *The Scotsman*, 13/11/86).

In early January a serious riot broke out in Barlinnie Prison in Glasgow. A number of staff were taken hostage and held over a four-day period. Again there was massive damage. Again the media had a grandstand view of what was happening (*The Scotsman*, 9/1/87, *The Sunday Times*, 11/1/87).

After the Barlinnie riot it became clear to the authorities that traditional methods of re-asserting staff control were insufficient for dealing with this new pattern of incidents. In the longer term, accommodation areas would have to be re-designed to prevent prisoners from rioting in large numbers and taking over large areas of a prison. In the shorter term, a more radical strategy would have to be developed to cope with further instances of this new phenomenon.

A New Response

There are three prison services in the United Kingdom, those in England and Wales, in Scotland and in Northern Ireland. Although they are organisationally independent, the two smaller services sometimes make use of the broader level of facilities available in England and Wales, such as training for senior staff. During the 1980s the Prison Service College in Wakefield had developed a command course for the management of serious incidents, which most senior governors in the Scottish Prison Service had attended. This concentrated on hostage incidents that took place in confined areas. Those taking part, were tested in the skills of command, of negotiation and of intervention. Over a period of five days participants were required to work on a series of increasingly sophisticated exercises, each of which developed according to how those, taking part responded.

The pressure on commanders and negotiators was stepped up as the complexity of the various scenarios increased. The responses were not always as expected. One exercise, which was scheduled to last for 24 hours ended after only an hour or so when the commander lost concentration and wrongly decided to send in an intervention team without checking the physical layout of the scene of the incident. By the time the team had negotiated the barriers set up by the prisoners the umpires decided that the hostage-takers had killed the hostages.

This whole structure was based on the premise that the local governor and his or her staff would manage any incident of this kind. When necessary, additional staff would be drafted in to boost the negotiating and intervention teams. If an incident went on for an extended period, a limited number of senior staff might be brought in to provide relief for the local management. This was the pattern, which was used in the incidents in Edinburgh, Peterhead and Barlinnie. This structure made a number of assumptions. The first was that all senior staff in every prison would be trained in incident command. Logistically this was not realistic since the SPS depended on an allocation of places from the Wakefield Staff College and this was never enough to satisfy demand. At the time of these incidents a relatively small number of Scottish prison governors had undergone this training. The second assumption was that all senior staff would make themselves available for training. At that time in the prison service there was reluctance among senior governors to undertake training of any kind. Many of them took the view that, from years of experience, they knew all that there was to know about their job, including how to deal with unexpected incidents such as these major riots. This meant that the management of the first three incidents described above was at times, to put it kindly, inconsistent and at other times downright dangerous.

In early 1987 the SPS conducted a thorough review of the management of these three major incidents. The outcome was a decision to set up an entirely new process for managing major incidents. Two complete incident command teams were to be set up to operate on a national basis. When a major incident broke out in any prison, a decision would be made by senior operational staff in prison service headquarters as to whether to call out these teams. If that happened, the teams would take over the management of the incident from the local governor and his staff. A Commander led each team and had units trained in negotiation and intervention techniques. In addition, an Operations Liaison Officer was to attend to act as a direct link when needed with the central command team in headquarters. This was a sensitive position as the person had no command authority but did have a role in reporting to the central command if he was of the opinion that the Incident Commander was not coping well with his responsibilities.

Before this radical new structure was set up there was consultation with, the Association of Chief Police Officers (Scotland) (ACPO(S)) Fire Services, the Armed Forces and other agencies. The new structure also required a senior police officer, usually of at least Assistant Chief Constable rank, to be present in the command room and for police staff to shadow the other prison service command teams. This was in recognition of the fact that a major incident, such as a riot or a hostage taking, was a criminal offence and as such came under the ultimate jurisdiction of the relevant Chief Constable. The prison service recognised that the police had a right to take over command of these incidents even though they were on prison property. However, ACPO(S) indicated that it did not wish police to take over command of these incidents except when the prison service command structure was unable to continue. For example, if it were decided that firearms would have to be used, the prison service Incident Commander would hand over command of the incident to the police. The senior police officer was, therefore, in the command room to advise the Incident Commander and available to take over after consultation if that proved necessary.

A wide selection of prison staff of varying ranks was invited to attend what turned out to be a demanding selection process at the Fire Service College in East Lothian at the beginning of April 1987. Most of them had previously taken part in the Wakefield command course. After a series of realistic exercises and assessments the composition of the two teams was decided. I was asked to be Commander of the first of the teams. The establishment of what was to become elite incident command teams met with a mixed reaction in the prison service, particularly among senior governors. Some of them regarded the plans to bring in outside teams to manage incidents in their prisons as a negative reflection on their own professional capabilities. Insult was added to injury by the fact that the other commander and I were at that time youngish middle ranking governors, even though in charge of our own prisons. Serious incidents tended to take place in the bigger prisons and we would therefore be coming in to take over from more senior colleagues. After the selection process was complete, arrangements were put in hand for further training but before that could begin fate took a hand. Hardly one week after the initial selection process was completed, there was another hostage incident at Perth Prison and the call went out to the new command teams to report for duty. The Alpha Team, which I commanded, was the first to be called to the scene.

Into Action

It is hardly surprising that, after a quarter century of experience at senior operational level in the prison services of the United Kingdom, one or two dramatic incidents are seared forever in the memory (Coyle, 1994).   I remember, for example, the morning in Greenock Prison in February 1987 when I arrived at the prison to be told that 230 prisoners were sitting in the dining room refusing to move because they had just heard on the radio that they were all to be moved to other prisons. The management of that incident tested all my negotiating skills. I also remember a different form of incident command when I was Governor of Brixton Prison in London. One lunchtime I was summoned urgently to a cell where I found a young man lying dead, having hung himself from the window bars. Another incident which I shall never forget was what came to be known as 'Perth 1'.

'Perth 1': April 1987

When our Incident Command Team arrived in Perth Prison we discovered that two prisoners were holding another prisoner hostage in one of the main accommodation halls. There had been a considerable amount of damage in the early part of the incident and other prisoners had been removed from the hall. In the course of the first few hours we set up the command team unit and after a settling in period we began to feel that we were making good progress. The negotiators had established contact with the hostage takers. There was some doubt as to whether the third prisoner was indeed a hostage or whether he was involved with the other prisoners. The intervention teams were on stand-by but were not expected to be needed. Halfway through the second day events began to take a much more worrying course. The prisoners became much more irrational. There was a real fear that the whole situation was going to escalate and that there would be serious injury to the hostage.

In the course of the night I had regular briefings with the leader of the negotiating teams and with the leader of the intervention teams. The tactics in situations such as this are always aimed at achieving a negotiated peaceful end to the incident. At the same time, the possibility of forcible intervention cannot be ruled out. With the other senior members of the command team, I went to a quiet corner of the command suite to assess the risks of intervention and of non-intervention. The risks were clear. Some of the intervention team would have to go in over the roof of the hall. Staff were not trained to operate at heights such as this where one false step could send an officer tumbling to his death. Even if staff worked very quietly and at great speed, the prisoners were likely to be alerted to what was about to happen. In their unstable condition there was a danger that they would panic and do serious injury either to their hostage or to themselves. On the other hand, it was their emotional instability that was causing us most concern. The negotiators were not building up any rapport with them and there was a distinct possibility that they would begin to take an even more violent course of action. I listened to all the arguments being rehearsed but knew that eventually the decision would have to be mine alone. I finally concluded that the only solution was to send in the intervention team to bring the incident to an end by force. I advised the Central Incident Command Unit in prison service headquarters and was given authority to proceed in my own time.

The intervention team leader presented me with his proposed plan of action. There were several entrances to the hall at different levels. He proposed that separate teams would go in at each level. The prisoners were in the upper part of the hall and there was a distinct possibility that when they heard the staff coming in that they would head for the extensive roof spaces and on to the roof, where they had been for part of the previous day. To prevent this, the intervention team leader wanted to lead a further team in over the roof of the hall. This would be a highly dangerous manoeuvre, involving staff gaining access to the external roof by means of a central tower and then making their way across the steeply pitched roof to access points. The intervention team consisted of fit young prison officers but they had not been trained to work at this height. It would not take much more than a lost footing to send one of them tumbling to certain death. I met the team members and was impressed by their confidence in the team leader and their willingness to carry out the proposed plan of action. I decided that the teams would go in at 5.30am, just before the rest of the prison came to life and when there was sufficient light for them to see what they were doing.

Responsibility of the Incident Commander

As the time approached I sat in the commander's chair and thought about what I was asking these staff to do. It was something far in excess of what they must have expected when they joined the prison service. There was great media interest in the hostage taking and everything that happened would be subject to minute political scrutiny. There was no way of being sure how the other 600 prisoners would react to the intervention. I had authority for what I proposed to do but, in the final analysis, the decision was mine. I could not be sure of getting all the credit if it went well. I would certainly be held fully responsible if it went wrong. The prison's closed circuit television cameras were relayed to the command room. I could see the high pitch of the roof across which the team would have to scramble become clearer as the dawn broke and was very conscious of the dangerous task which I was asking the young officers to carry out.

Shortly after five o'clock the team leader reported that his men were all in place and he was waiting for the coded signal to intervene. I took my staff officer aside and we spoke quietly for a few moments to satisfy ourselves that everything was in place. I then returned to the commander's chair and gave the coded signal to the communications officer. I heard him bark it out through the Cougar radio. Then there was silence. There were six or eight other people in the command room. We had trained together as a team for eventualities such as these and we had built up a close bond in the course of this incident. But suddenly it felt as if the team did not exist. The responsibility of command was in a single line, which ended with me. Having exercised that command, matters were now out of my hands.

The tension in the command room was palpable as we watched the television screens. We could see shadowy figures moving across the rooftop and disappear into the roof spaces. The silence of the action meant that we were not in a position to maintain radio contact. I had to trust the intervention team to do what was right. After a few moments we saw what looked like puff of smoke coming through the slates. There was a collective gasp in the room and I gripped the edges of my desk. Was this to be the realisation of my worst fears and the beginning of a major human catastrophe? After what seemed an eternity the radio crackled into life: 'Location secure. Everyone safely accounted for.'

After that there was a frenzy of activity. The prisoners were taken out quickly for medical examination and debriefing. The hall was given a thorough security sweep to ensure that no booby traps had been left. The intervention teams were debriefed and praised for their work. It was equally important that the negotiation teams were also debriefed and re-assured that they should not see the need to use intervention as a reflection of their failure. Full reports had also to go back to the central incident command unit in Edinburgh. It was only much later as I was driving home that I had time to go over in my own mind what had just happened and the extent to which my decisions could have been a matter of life or death for many of the men involved. For the first time I understood what was meant by the loneliness of the incident commander.

In many respects the role of Incident Commander reflects what a prison governor has to do in the course of normal duties. I well remember my first day as Governor in charge of Greenock Prison, sitting in the governor's chair at the morning meeting and suddenly being very conscious that everyone was looking towards me. Whatever I said on any topic would become the management decision. At the same time, one had to create opportunities for other people to feel confident enough to voice their own opinions and to contribute the benefit of their experiences. They were generally happy to do so, knowing that when the discussion was over 'the old man' would decide what was to happen (Miller, 1976). In principle, the same applies in the command of a serious incident. One can create an environment in which there is room for sharing of ideas and discussing various possible plans of action. In my case, I did this by having a group of chairs round a coffee table at one end of the room. This was the discussion area where options were considered and everyone had a voice. When the time came for decisions, I returned to the commander's chair and everyone knew that we were in command mode.

At another level, of course, the leadership demanded of the incident commander was of a different calibre from that which had to be exercised by a governor on a daily basis. Daily decisions were generally managerial. They might have long term operational implications but they did not normally have an immediate consequence (Grew, 1958). Those taken by an incident commander frequently did; whether it was the decision to send an intervention team in over rooftops or to instruct negotiators that they should not concede a demand being made by an emotionally unstable hostage taker.

Evaluating the New Approach

This first experience of the new incident command structure followed a text book pattern although it is true to say that in many respects we were writing the text book as the incident unfolded. This model of incident management was already being developed in emergency services but this was the first time that anything of the sort had been used in a prison setting. In many respects we were fortunate that the incident occurred in Perth Prison. This was the fourth of what at that time were the major prisons in Scotland (Cameron, 1983). It was the only one in that group not to have suffered a major incident in the previous six months. It also had a good reputation for the general relationship between staff and prisoners. Most importantly for us, the governor was at ease with the new command structure. He gave me and the rest of my command team a full briefing as soon as we arrived and then handed over the incident area to my command while he retained authority for the rest of the prison. Throughout the incident one of his senior staff was nominated as the liaison person between the resident prison management and the incident command team and fulfilled our every request.

In preparing for incidents of major unrest in the prison service the first aim was the safe release of the hostages. Linked to this was the objective of reaching a peaceful conclusion through negotiation rather than intervention. The possibility that intervention might eventually be required, had always to be borne in mind and it was significant that this first outing for the new incident command teams ended with intervention rather than by peaceful negotiation. With the benefit of hindsight, it is possible to see that this was not necessarily a bad thing. Many staff in the service had been shaken by the fiasco of the bungled attempts at intervention during the Barlinnie riot in January 1987 and the disciplined intervention in Perth showed that such action could be managed safely and successfully, without injury to either staff or prisoners. It also sent a message to prisoners who might be tempted to be involved in dangerous incidents like this that incidents could be brought to a violent but disciplined conclusion on staff terms. In the immediate aftermath of what was to become known as 'Perth 1' and in the subsequent debriefs which took place, it was emphasised time and again that intervention of this nature was not an adverse reflection on the efforts of the negotiating teams, rather that everyone had to recognise the part they played in what was a highly disciplined team effort.

The greatest importance of this incident in management terms was that it established the value of the concept of centrally organised incident command teams. In this and subsequent incidents the team itself was bonded in a way which I had never experienced before or since in the prison service. At the same time the team worked closely with local staff in the prison. I made sure that after the event the prison governor was with me when we went to congratulate staff. The fact that the Intervention Team commander happened to be a Principal Officer from Perth was a distinct advantage.

The development of this form of specialised incident command, in which the best people were selected for the relevant positions and given appropriate professional training, had far-reaching consequences for the general management of the SPS. The out-dated principle that people should be promoted to the most senior ranks solely by virtue of length of service and once there, were thought to be the founts of all wisdom was seen to be no longer acceptable. The governors who thought they could manage major incidents by sheer force of personality in a manner which reflected their idiosyncratic way of governing their prisons were found to be lacking and, most important of all, were seen to be putting lives, whether of staff or of prisoners, at risk. Within a year or so there was a complete change of senior governors in the SPS.

Development of the Incident Command Teams

After 'Perth 1' there was a period of uneasy calm in the Scottish Prison Service. What many of us suspected then and was subsequently proved to be the case was that this was the prelude to much more serious disturbance. This was the period of 'Grand Design' described above, when there was massive re-allocation of prisoners serving long terms of imprisonment. When all of this movement had been completed with what appeared to be minimal disruption it was decided that the training of the new Incident Command Teams, which had been suspended in April 1987, should go ahead. In the middle of September the two recently formed Incident Command Teams went to Redford Army Barracks in Edinburgh for an extended training period when we were subjected to a series of mock incidents which matched and indeed exceeded anything that we might be expected to have to deal with in practice.

Shotts Prison: September, 1987

Shortly thereafter many of us began to believe in the theory of the self-fulfilling prophecy. The first training period ended on 16 September. Ten days later an officer was taken hostage in the modern prison for long term prisoners at Shotts in Lanarkshire and the Incident Command Teams were called out; again Alpha Team was to be first on the scene. By this time one had a real sense of symbiosis. On the one hand, there was distress at a personal level that another young officer had been placed in a life-threatening situation and there was anger that yet again the SPS was to be pilloried in the media for a failure to control its prisoners. This was balanced against a real surge of adrenaline that the professional skills, which one had developed in the course of training, were to be tested in a real life situation. At a very personal level, there was anticipation that one's skills as a leader of men were about to be tested to the limits.

In the event, this incident was again dealt with in textbook fashion. It lasted 25 hours, which meant that our Command Team was in charge at the beginning and at the end of the incident, a satisfying scenario in terms of procedures. On this occasion the incident concluded safely through a process of patient negotiation and the young officer who had been held at knifepoint was safely released (*The Glasgow Herald*, 28/9/87). The main feature of this incident was that it brought to official attention the fact that these incidents might not be sporadic acts of spontaneous violence but rather could be part of a much more organised sequence of events. Half way through the incident, the negotiators felt that they were beginning to make good progress in building up a rapport with the prisoners who were speaking for the others. In the command room, at one step removed from the negotiation, we felt that something was not ringing true. There was already evidence from earlier incidents that prisoners had grasped the principles on which our negotiation was structured and had built up a mirror structure. In Shotts we had the feeling that the prisoners we were negotiating with were not as confident or assertive as we might have expected them to be as leaders. There were also significant delays in getting responses from them to our proposals. At one point a large group of the prisoners who were not involved in the negotiations began to bang and shout. Suddenly a voice rang out telling them to be quiet and an immediate sudden silence fell. In the command room we are alert. This was a voice we had not heard before. It was not one of the prisoners on our list. We replayed the tape many times and eventually one of the local staff was able to identify the voice as belonging to a prisoner who was serving a long sentence but who had never before come to the notice of the authorities. Once he had been identified we were able to confirm that he was one of a family of well-known Glasgow criminals, who were at the time involved in a battle with another gang for control of the drug scene in that city. Subsequent events confirmed the truth of this evidence. Armed with this information we were able to change our negotiating tactics and this undoubtedly contributed to the speedy end to the incident.

Hostages at Peterhead: October, 1987

This Shotts incident was the beginning of two weeks of disruption, which almost brought the SPS to its knees. The day after the Shotts incident ended, two officers were taken hostage in Peterhead in a riot that lasted for five days. The day after the conclusion of that incident, an officer and a number of prisoners were taken hostage in Perth Prison. The Peterhead riot broke out just as the two Incident Command Teams were beginning to unwind after their exertions at Shotts. A decision had to be made as to whether individuals were emotionally and psychologically fit to go straight into a new incident, which clearly had the potential to be much more traumatic than that at Shotts. The prisoners at Peterhead had already been classified as the most troublesome in the system and a number of those involved in this latest incident had been involved in previous riots and had been transferred to Peterhead for that reason. The two officer hostages were beaten at an early stage in the incident and one of them had a broken leg. This incident also brought to a head the simmering disagreement between the central prison authorities, who had developed the concept of national Incident Command Teams, and the number of senior governors who took the traditional view that they should be left to resolve any trouble which arose in their own prisons. That was very much the view of senior management at Peterhead, none of whom were involved in the national Command Teams.

The compromise, which was reached, was an unhappy one, which may have contributed to the further violence that ensued. It was decided that local senior management should retain control of the incident with the governor acting as Commander. However, middle and junior members of the national Incident Command Teams were sent to Peterhead to act under them. The two national Commanders were unhappy to see their teams being dispersed in this way and the team members themselves felt unsure as they were sent into a situation in which they were expected to operate the principles of the new way of working without the structure which had already proved so successful elsewhere. The management of the incident was further complicated by the fact that the command room in Peterhead had large windows, which gave a very clear view of the roof of the cellblock where the incident was taking place. Looking out of these windows the commanders could see the officer hostage being dragged over the high roof with a chain around his neck and the perpetrators threatening to throw him to his death. This riot ended with violent intervention and the release of the officer (*The Daily Record*, 1/10/87, *The Scotsman*, 5/10/87). Information provided at later court proceedings confirmed what was suspected at the time, that the successful intervention team was made up of members of the S.A.S. (*The Observer*, 26/6/88 and 19/2/89).

Perth 2: October, 1987

When a further incident broke out the following day, 4 October 1987, in Perth Prison the full incident command teams were mobilised. At that juncture the whole of the SPS was traumatised by what seemed to be an unstoppable sequence of major riots. The terrible events on the roof of Peterhead Prison had been recorded on camera by the world's media. The dramatic end to that incident, which had also been recorded by the media, seemed to call into question the validity of the Incident Command strategy. Given this background, the incident that then unfolded in Perth was the most demanding in which we had been involved. Terrible as the incident turned out to be, the Command teams were at one level glad to have the opportunity to be back in action so quickly to demonstrate that the SPS had learned how to manage serious riots. However, we were also aware that the violent end to the riot at Peterhead had raised the stakes considerably. Staff throughout the Service felt that it was only a matter of time before someone was killed. This was reinforced when prisoners standing at windows began to stage mock executions and produced banners proclaiming 'This siege will end in death.' (*The Scotsman*, 5/10/87 and 6/10/87).

By the time of the outbreak of this new incident in Perth we were also becoming used to the need to make use of our experience of previous incidents. This was the third such outbreak in a Scottish prison within as many weekends. The full teams had been involved in the first and the third. The majority of the team members had also been involved in the second one, in Peterhead. The commanders, who had not been involved, had watched from afar and were keen to get back into operation with their teams. We were also conscious that we were becoming more skilled with each incident. Without becoming blasé, we could almost predict what was likely to happen at the 12, 24 and 48-hour stages. This was an advantage that we had over the perpetrators. For all of them, each incident was their one and only such experience.

I was reminded of this fact shortly after I was appointed as Governor of Peterhead Prison in April 1988. As part of the process of restoring order in the Scottish Prison Service it had been decided that all prisoners who had been involved in any rioting, taking hostages or other serious disturbances would be held in Peterhead. When I took command there I found myself in daily contact with all of the prisoners whom I had come to know intimately for a few short but intense days during each of the incidents. The grapevine of the Scottish Prison Service is legendary. No confidence can be kept for very long. For some reason, the Service was able to maintain complete confidentiality about the roles of individual members of Command Teams. Even when the media carried reports about these new 'elite squads' they were unable to name any individual member (*The Scotsman*, 11/8/88). Almost unbelievably, the same applied to the prisoners. I remember listening to one of them telling me in no uncertain terms what he thought about the man who had been commander during the incident in which he had taken a hostage and informing me in graphic detail what he would like to do to him. Little did he know that this rogue of uncertain parentage was standing in front of him.

Pressure from Politicians and the Media

Another new feature of the incident in Perth in October 1987 was the degree of involvement, not only from prison service headquarters, but also from the Scottish Office and from politicians. In previous incidents there had been close communication between the Incident Commander on site and the central control room in prison service headquarters. The latter had been a source of advice and of confirmation that the action being taken was appropriate but had made no attempt to influence operational decisions. This time the pressure was quite different. This was the third major breakdown of control in a Scottish prison in successive days. Prisons had been the top media story through all of this period; there were dramatic pictures in both press and television. And this was also the time of year for party political conferences. 'Perth 2' started on Sunday 4 October. The Conservative Party Conference was scheduled to begin on 6 October. The last thing that the Secretary of State wanted was to appear at the conference amidst pictures of chaos and anarchy in his prisons. Late on the second evening of the incident I was talking from my command chair to the senior operational person in the headquarters command room when he suddenly told me that the Secretary of the Home and Health Department was in the room and wished to speak to me. This was highly unusual and had not happened in any previous incident. At that time the Prison Service was organisationally part of the Home and Health Department and the Director of the Service was responsible to the Secretary of the Department. He was the most senior official but should have had no involvement at an operational level. I assumed that he was in the central control room to demonstrate his support for staff there and simply wished to speak to me to pass on his encouragement to the Command Team. Instead I found myself being closely questioned about my actions and my proposals for ending the incident speedily. I suddenly sensed that the Secretary of State himself was also present in the headquarters command room. This was a none-too-subtle pressure, which a Commander could well do without.

This aspect of incident command had not been dealt with in any depth in our training sessions. It was recognised that no incident command team could act in a vacuum. In the first place there had to be close collaboration between the two teams likely to be involved in managing any incident on a continuous basis. There also had to be close liaison between the incident commander on the ground and the Central Incident Command Unit in prison headquarters. In this respect the Operational Liaison Officer played a key role. From one perspective, he was the spy in the camp. His formal role was to act as liaison and to make sure that the lines of communication with the Central Command Unit ran smoothly. He also acted as a middleman between headquarters and the governor who was still managing the rest of the prison. But he also had an informal role, which was to report back to senior staff at headquarters about how the commander was performing, about how he was coping with the stress of command and about whether he was demonstrating the necessary qualities of command. At that time we were at the cutting edge of that type of incident command. No-one could be sure how commanders might react to the pressures of these entirely new situations.

In reality, all these relationships worked well because there was trust between the various actors. To a large extent it was important that we were all committed to this new way of working. Others were waiting to see it go wrong and those of us involved, at whatever level, were determined to pull together so that it would succeed. As this form of managing serious acts of unrest in prisons developed elsewhere there were examples of negative tension between local and central command, between what came to be known as Silver and Gold Commanders. The best known example of this occurred in the Strangeways riot in April 1990 when the local commander wished to intervene forcibly at an early stage and the headquarters commander prevented him from doing so (Home Office, 1991). This thankfully was never an issue in Scotland in my experience. There was an acceptance on the part of the Central Command Unit that ultimately the local Incident Commander had to be allowed to make his decisions. To balance that out, the local commander invariably talked major decisions through with people in the Central Command Room. It was, of course, easier to work in this way in an age when communication was less immediate than it is now. In times of need, a broken telephone link could become the modern equivalent of the telescope put to Nelson's blind eye.

A very important role for those in the Central Command Room was to shield the Incident Commander from political and other external pressures. The only time in my experience when this almost broke down was on that occasion immediately before the 1987 Conservative Party conference.

A Peaceful Conclusion

Another feature of this riot in Perth was the instability of the two main prisoner perpetrators. They frequently threatened to inject heroin and infected blood into the officer who was being held hostage (*The Scotsman*, 11/8/88). They captured the prisoners in the hall who had been convicted of sexual crimes and began to stab and cut them. At one point negotiators reported that a severed finger had been thrown out, although this was not in fact the case. Serious consideration was given to ending the incident by intervention. The manner in which the recent riot at Peterhead had been resolved made me more reluctant than usual to decide on this option, particularly because of the danger that the unstable perpetrators would respond violently at the first sign of intervention; there was a real danger that someone would be killed. With their knowledge of what had happened at Peterhead, the Tayside Police were equally reluctant to become involved in violent intervention. As we were considering various options, the Assistant Chief Constable warned me, 'I hope you are not expecting my men to go abseiling from the roof the way they did in Peterhead. Remember we're only Tayside Bobbies!' Finally, with the involvement of the prisoners' lawyers we were able to negotiate a safe conclusion, although on this occasion the officer was very badly shaken and a number of the prisoner hostages were seriously scarred. This successful and relatively peaceful outcome went a long way towards restoring the professional confidence of staff throughout the Scottish Prison Service. It also confirmed the wisdom of using highly trained command teams to deal with this sort of incident.

The incident in Perth in October 1987 was the last in that immediate series. Had there been even one more soon after that, the whole structure of the Scottish Prison Service might well have been unable to withstand the pressure from within and without (*The Times*, 12/8/88). In the event, the original Incident Command Teams had to deal with a third major incident in Perth Prison in May 1988. In this an officer was taken hostage and sex offenders were again severely beaten. The incident ended by negotiation after 18 hours (*The Scotsman*, 11/3/89 and 14/3/89). By this time additional Command Teams had been identified and were undergoing training. Radical changes were made to senior personnel and to management strategies in the Scottish Prison Service which led to a new sense of direction which developed in the following years.

Conclusions

Within a few short months between 1987 and 1988 the Scottish Prison Service made a quantum leap in the way it managed major incidents of unrest among prisoners. There are at least three conclusions to be drawn from our experience at that time.

- The separation of command of serious incidents from normal daily management skills was an entirely new concept for the prison service. It was only much later, as a result of international interest from other prison services, that we became aware of the extent to which we were breaking new ground in the Scottish Prison Service in 1987 and 1988. This principle is now generally accepted (Flin, 1996).

- Faced with the absolute need to train the best people for the task in hand, the Scottish Prison Service for the first time abandoned its tradition of selecting people according to seniority. This process of assessment, selection and training of the best people later found its way into more general application and changed the whole senior management structure of the Service. This in turn led to a period of radical thinking about the way the Service should be managed and about the treatment of prisoners (Scottish Prison Service, 1990). The period of relative calm, which followed, can be attributed to this fundamental reform.

- These changes were made possible because of a determination among senior Prison Service staff that the service should become more professional in its way of working. William McVey, the newly appointed Controller of Operations, was a key figure in driving through these changes (*The Scotsman*, 11/8/88). It is a feature of progressive change in many prison services that one or more individuals have crucial roles to play.

Lessons for the Future

- An important feature of this more professional approach was the close co-operation between the Prison Service command structure and that of the emergency services, principally the Police, but also the Fire and Ambulance Services. The strong professional friendships, which were established, spread into wider co-operation in other pertinent matters (Scottish Prison Service, 1987).

- There was also a recognition that hostage incidents in prisons needed to be managed differently in several respects from other hostage incidents usually dealt with by the police. One clear example of this difference was in negotiation tactics. Generally speaking, police negotiators are given delegated authority to use their initiative in certain circumstances in respect of what and when to concede to limited demands. Hostage incidents dealt with by the police are usually single incidents and decisions made during them are unlikely to affect any similar incidents elsewhere. This was not the case with the hostage incidents in prisons. As the series of incidents unfolded we were very conscious that perpetrators knew what had been conceded in previous incidents and wanted to use that as a given at the start of negotiations. As negotiations continued in one incident, Commanders were very conscious that decisions would affect the management of any future incident. In a wider context, Commanders were also aware that the prison would have to return to normal after the incident had ended and that decisions or concessions made in the course of the incident might make the local Governor's life more difficult after the Command Team had packed its bags and departed. This principle was accepted by the prison service negotiators as being to their advantage. It was easier for them to accept this than for police colleagues since, even in normal prison life, one way for an officer to cope with an angry prisoner is to explain that his hands are tied and he has to obey orders.

- One of the most important elements of a Commander's work was to find the balance between making sure that other team members were fully consulted and that everyone was able to make his or her contribution while at the same time making sure that people understood the importance of implementing command decisions once they had been made. The ploy that I used for this was to have a small group of chairs at one end of the command room itself. When I was consulting I sat in one of these chairs with other staff gathered around. Once the consultation was complete and a decision needed to be made I returned to the large chair at the Commander's table. Other members of the team said that they found that a very useful pointer for them in recognising when I was issuing orders. To some extent this ploy underlined the importance of symbolism in management in general and incident command in particular. The people involved in unusual and stressful incidents need to have clear reference points. Commanders have to be aware of this need and to provide them at appropriate points.

Note

1. International Centre for Prison Studies, King's College London, 75–79 York Road, London SE1 7AW, United Kingdom.

References

Cameron, J. (1983). *Prisons and Punishment in Scotland*. Edinburgh: Canongate.
Coyle, A. (1994). *The Prisons We Deserve*. London: Harper Collins.
Coyle, A. (1991). *Inside: Rethinking Scotland's Prisons*. Edinburgh: Scottish Child.
Cressey, D.R. (1961). *The Prison: Studies in Institutional Organisation and Change*. New York: Holt, Rinehart and Winston.
The Daily Record. (1/10/87). 'Hammer of Hate.'
Flin, R. (1996). *Sitting in the Hot Seat: Leaders and Teams for Critical Incident Management*. Chichester: Wiley.
The Glasgow Herald. (13/11/86). 'Turning the Security Screw a Notch Tighter.'
The Glasgow Herald. (28/9/87). 'Prison Officer is Freed as Jail Protesters Surrender.'
Grew, B.D. (1958). *Prison Governor*. London: Herbert Jenkins.
Home Office (1991). *Prison Disturbances April 1990, Report of an Inquiry by the Rt. Hon. Lord Justice Woolf (Parts I and II) and his Honour Judge Stephen Tumim (Part II)*, p 76–79. London: HMSO.
Miller, A. (1976). *Inside Outside: The Story of a Prison Governor*. London: Queensgate Press.
The Observer. (26/6/88). 'Crisis in Scottish Jails as Violence Continues.'
The Observer. (19/2/89). 'Legal Fury Over Man in SAS Jail Siege Case.'
The Scotsman. (6/10/86). 'Prisoners Stage Mock Killings.'
The Scotsman. (30/10/86). 'Hostage Officer 'Bearing Up Well.''
The Scotsman. (13/11/86). 'Five Prisoners in Siege Swap.'
The Scotsman. (9/1/87). 'Hostage Freed in Jail Siege.'
The Scotsman. (5/10/87). 'New Jail Hostage Crisis. Four Prisoners Seize Officer at Perth.'
The Scotsman. (5/10/87). 'Prison Officers Demand Tighter Control on Hard Men.'
The Scotsman. (11/8/88). 'The Riot Squads Go Into Action.'
The Scotsman. (11/8/88). 'We'll Blow His Mind and Give Him AIDS.'
The Scotsman. (11/3/89). 'Prison Officer Tells Court of Accused's Death Countdown.'
The Scotsman. (14/3/89). 'Riot Trial Told of Threat to Cut Off Prisoner's Leg.'
Scottish Prison Service. (1987). *Report of the Review of the Handling of the Three Major Incidents in Scottish Prisons, October 1986 to January 1987*. (Unpublished).
Scottish Prison Service. (1990). *Opportunity and Responsibility: Developing New Approaches to the Management of the Long Term Prison System in Scotland*. Edinburgh: The Scottish Office.
Scottish Prison Service. (1998–99). *Scottish Prison Service Annual Report and Accounts*. Edinburgh: The Stationery Office.
The Sunday Times. (11/1/87). 'Defiance Song Ends Siege of the Big Bar-L.'
The Times. (12/8/88). 'The Prison under a Sentence.'

Wozniak, E. (1989) *Current Issues in Scottish Prisons: Systems of Accountability and Regimes for Difficult Prisoners.* Scottish Prison Service Occasional Paper No.2. Edinburgh: The Scottish Office.

Part III
Developing Commanders and their Teams

11 Command Decision Making

Margaret Crichton and Rhona Flin
Industrial Psychology Group, University of Aberdeen

Margaret Crichton, MA, MSc, is a Research Fellow at the University of Aberdeen, examining training for emergency response and command decision making. She is currently completing her PhD research on decision making in emergencies.

Rhona Flin BSc, PhD, CPsychol, FBPsS, FRSE is Professor of Applied Psychology at the University of Aberdeen and Director of the Industrial Psychology Group. She conducts research into industrial safety, Crew Resource Management, emergency management, and incident command. Her books include 'Sitting in the Hot Seat' (Wiley) and 'Decision Making Under Stress' (Ashgate: edited with Salas, Strub & Martin).

Introduction

Command decision making continues to be a crucial topic for any organisation concerned with managing major incidents. Crises (natural or man-made), civil turbulences or terrorist actions, according to Rosenthal and Pijnenburg (1991), can be characterised by "un-ness – unexpected, unscheduled, unplanned, unprecedented, and definitely unpleasant" (p. 1). For those called upon to manage the crisis, such circumstances normally engender a change in management style from consultative to 'command and control'. While there is a voluminous literature on military command (see Larken, Keeling, this volume) and crisis management at a political (Stern, 1999) or strategic level (Rosenthal, Comfort & Boin, in press), relatively little has been written on the skills required for emergency management at the tactical or operational level, in a non-military setting. In

201

the last decade, a number of on-scene commanders (both civilian and military) have been found to have made decision errors at demanding, high risk incidents – e.g. Hillsborough (Taylor, 1990), Piper Alpha (Cullen, 1990), and the US Vincennes (Fogarty, 1988). As a result of a series of high-profile failures in command decision making, both military and industrial psychologists (Flin, 1996; McCann & Pigeau, 2000) have begun to examine the skills required for effective incident command.

This chapter will present a selective review of current psychological research relating to incident command, especially focussing on decision making, one of the essential command skills. The first section discusses the features of incident command, particularly at the operational and tactical levels. This is followed by an overview of recent advances in decision research, with a discussion of decision strategies, situation awareness, and the impact of stress. The chapter ends with a look at methods for training incident commanders, especially low-fidelity techniques for improving command decision making.

What is Incident Command?

Command, according to the emergency services, involves the exercise of authority to meet the responsibilities placed upon the Incident Commander, which may include legislative, contractual, political and moral responsibilities (Home Office, 1997). Moreover, it refers to the act of directing and controlling personnel and resources within an effective operational plan based on the available information. Command requires high level decision making skills, normally acquired from experience. Control, on the other hand, is more of a monitoring function, which relies on the gathering and processing of information: it lends itself to proceduralisation and checklists. Control relies on feedback that allows the commander to adjust and modify command action as required (Pigeau & McCann, 2000; US Marine Corps, 1996). Command and control is therefore an interactive process involving all parts of the emergency response organisation. The incident commander does not and should not work alone, but is part of a command team. Effective incident command therefore relies upon the knowledge, skills and attitudes of both the command unit team (see Salas, this volume) and the incident commander.

Command and control skills have been a pre-requisite for senior personnel in the military and the emergency services for as long as these domains have existed (see Brunacini, 1985; Cannon-Bowers & Salas, 1998; McCann & Pigeau, 2000). In recent years, it has been acknowledged that a similar set of command skills is essential for key decision makers/leaders in industrial emergency response organisations (ERO) (e.g. Channel Tunnel Safety Authority, 1997; offshore oil installations – Cullen, 1990, nuclear power plants – Crichton et al, 2000a), where the potential for serious accidents also exists with risks of fatalities and environmental damage.

Incident Command Structure

Within the UK emergency services (and adopted by many industrial ERO), the command structure is based on the strategic/ tactical/ operational levels described in "Dealing with Disaster" (Home Office, 1997; see Arbuthnot, this volume). In brief, incident management, at a strategic level, can take the form of two discrete patterns, dependent upon the nature of the tactical level. In a planned operation, the strategic level establishes the framework within which tactical commanders work, and occurs most often in military and police operations where there is pre-planning and 'pro-activity'. When an event is unplanned, and where the command is reactive, at least initially, strategic command has to start by embracing what tactical command has already established, offering support and facilitation. At the strategic level, co-ordination of multiple sources of information and management of resources is required. Strategic command is generally located away from the incident scene. Tactical command occurs at an Incident Command Point, located close to the incident scene, and is concerned with the prioritisation and allocation of resources, as well as planning and co-ordination of the response. The operational level directs and carries out immediate response actions at the incident site(s). The aim of a standardised, hierarchical, command structure is to improve communication, and minimise inter-agency confusion, in order to achieve more effective overall incident management.

The adoption of a command structure by ERO in industrial settings is also beneficial to effective incident command, as the ERO generally consists of ad-hoc teams who are only assembled in response to emergency situations. Members of these ERO do not normally work together in the same way on a day-to-day operational basis, making a standardised structure with clearly defined command roles and responsibilities, even more important. In an industrial ERO, as in the emergency services, the management of the incident depends upon the skills of those in operational and tactical command positions, but these will typically be on-duty managers and team leaders, rather than specialist incident commanders. This chapter will concentrate primarily on the tactical and operational command roles, where decisions must be taken in uncertain, stressful, complex, and time-pressured conditions.

Command Skills

The adoption of a formal command structure is not in itself sufficient to guarantee proficient command and control of an incident. Effective performance by the commander, and indeed the whole command team, is crucial. The skills of the commander will determine the effectiveness of incident management. According to Flin (1996), FSAB (1995), London Fire Brigade (1995), OPITO (1997), Pigeau & McCann (2000) and Sarna (this volume), these skills are generally deemed to comprise the following:

- Decision making
- Situation assessment
- Leadership
- Planning
- Communicating
- Monitoring
- Delegating
- Prioritising
- Team coordination
- Stress management.

Indeed, the very minimum for battle command, according to the US military, consists of leadership and decision making (Fallesen, 2000). In many high reliability work domains (aviation, medicine, energy sector industries), there is now a particular emphasis on the training and assessment of these cognitive and social skills, which are regarded as key to safe and efficient operational performance. In aviation these are called non-technical skills (O'Connor et al, in press; van Avermaete & Kruijsen, 1998) or Crew Resource Management skills (Helmreich & Foushee, 1993), and they will be discussed later in the chapter in the section on training.

It must always be borne in mind that the conditions under which command skills are exercised are generally demanding, risky and uncertain. To illustrate those conditions, and a commander's use of these skills, we have selected the incident aboard the HMAS Westralia in Australia in 1998. This highlights one of the worst decisions a Commander can face – whether to sacrifice the few, to save the many.

Incident aboard the HMAS Westralia, 5th May, 1998

An explosion aboard the HMAS Westralia, a 40,000 tonne Royal Australian Navy supply tanker carrying 20,000 tonnes of diesel fuel, resulted in Commander Stuart Dietrich being faced with not one, but two critical, life and death decisions. His ship and crew were in danger unless a blaze in the engine room was extinguished, yet four sailors (one woman and three men) were trapped in the engine room. Having just cleared Freemantle harbour, reports were received from the engine room of a leak on one of the ship's engines followed by a fire onboard. Due to the tone of the engineer's voice, Dietrich recognised that this was serious. His first thoughts were that the fire had to be dealt with, but simultaneously, he was also considering how commitments for the next day could be maintained.

"The next thing that happened was that all the fire alarms on the bridge went off and in fact all sorts of alarms went off, toxic gas, short circuit for the radio, fire alarms, every siren, or whistle, or bell that could ring, all started sounding at the same time."

The first action he ordered, in accordance with practice drills, was to deploy rescue teams to the locality of the blaze. Training exercises for this possibility had been specifically practised. This was not a difficult action as it was determined on the basis of familiarity and pattern recognition.

"You have the comfort of having done these drills before and so your body and your brain just keeps going with it. While it's not robotic, you have that as a sort of a crutch and it goes on. And so, the familiar patter is going on, coming in, and people are reporting in from all around the ship that they are ready from all round the ship.

And so that is going on, and in the meantime, there is 'Get that bloody alarm switched off' and 'Shut that door' – to keep the smoke out. As smoke was starting to come into the bridge through doors. So yes, it's hard to keep concentrating but it just flows the way you have practised before."

However, subsequent reports indicated that the fire was escalating rapidly, and the Engineering Officer could no longer take actions to control the fire. He recommended that, although initial reports of one missing sailor had since increased to four, the engine room should be drenched with carbon dioxide to extinguish the flames. [This action would suffocate any survivors]. Commander Dietrich made his first difficult decision, under time pressure. Against recommendations, he chose not to drench, but to send in a further fire fighting team. Appreciating that the fire was becoming more and more established and spreading, now threatening the safety of the ship and all her crew. Dietrich lived in the hope that the missing sailors might still be alive. Waiting for further information to be relayed back was agonising, with the distance between the bridge and the incident scene ever apparent.

"That was a frustrating and agonising wait because you are relying on information to be relayed back and the flow of information was hampered by the conditions in which these guys were working. At the same time you are high up on the bridge, you are dying to find out what is going on and hear if they are making progress and on top of that going through your mind is how much longer can I wait, because this fire is getting a greater and greater hold and I cannot let this go on for ever, because then we come down to the danger of it spreading and losing the whole ship."

Upon receiving word that the rescue teams had failed to locate the missing personnel, Commander Dietrich then had to make his second critical decision. He again recognised that the situation was constantly deteriorating and that the chances of survival for the missing sailors were minimal. He then decided to order the lethal carbon dioxide drench. Had there been any possibility of the survivors being found alive, he considered that his decision may have been different, however, on the basis of knowledge and experience, he felt that he had taken all possible actions. The safety of the ship and the remainder of the crew were his responsibility, and his

> decisions had to take this into account. In an official inquiry, the missing sailors were found to have died as a result of the toxic fumes of the fire, not the carbon dioxide drench, thus vindicating the Commander's second decision.

Source: The Times, 6.5.98; BBC Radio 4 , "The Choice" (23.2.99).

The HMAS Westralia accident shows clearly the time-pressured dynamic, high risk conditions in which commanders must take decisions, often without full information on the status of unfolding events. It demonstrates the enormous responsibility inherent in his moment of choice, but also reveals the significance of his experience and training. Until recently, very little scientific research had been carried out into the skills required to make decisions in hazardous work settings. The following section outlines the Naturalistic Decision Making approach, developed by psychologists to study experts making decisions in exactly these circumstances.

Naturalistic Decision Making

There is growing recognition that real-life, high-pressure environments place special demands on decision makers, particularly through heightened levels of stress. Classical, or rational decision making models may be less than adequate in offering viable explanations and applications for such domains. Identifying the most effective decision making processes used by experts is a prime objective for many organisations and agencies where decisions require to be made under stress. Our understanding of command decision making has been enhanced by new research into decision making in real-life environments, especially those characterised by high demand and risk. This field of decision making research is known as Naturalistic Decision Making (NDM) defined by Zsambok and Klein (1997) as:

> the way people use their experience to make decisions in field settings. (p. 4)

> how experienced people, working as individuals or groups in dynamic, uncertain, and often fast-paced environments, identify and assess their situation, make decisions and take actions whose consequences are meaningful to them and the larger organisation in which they operate. (p. 5)

The goal of NDM research is to examine the way people make decisions under time pressure, shifting conditions, unclear goals, degraded information, and team interactions (Klein, 1997a). In fact, NDM has three common themes, namely, the real-world context of the decision making, the importance of studying experts, and the focus on situational diagnosis as opposed to selection of action (Lipshitz, Klein, Orasanu, & Salas, in press). NDM research has been conducted in dynamic environments such as flight decks, military operations, firegrounds, medicine, and high hazard industries (Flin, Salas, Strub, & Martin, 1997; Zsambok, 1997). The essence of NDM is how a person uses experience and knowledge to tackle a problem, whereas traditional, analytical decision making often treats prior experience as a variable to be controlled, counterbalanced, or otherwise ignored (Pruitt, Cannon-Bowers, & Salas, 1997).

NDM has grown in appeal as a decision making approach due to its relevance to decision making in the real world (see Salas & Klein, 2001), though it is not without its critics (Yates, 2001). Rather than being prescriptive about how decisions *ought* to be made, NDM theories and models are descriptive in that they outline how decisions makers *actually* make decisions especially in time and risk pressured environments. Within NDM, new decision models have emerged, such as the decision process model (Orasanu & Fischer, 1997), and a model of recognition primed decision making (Klein, 1989; 1993), many of them influenced by earlier models of cognitive processes (Rasmussen, 1983). NDM models contain fundamental differences in their approach to the decision making process, as they tend to have been developed in the context of the domain in which the study has been conducted. So, for example, aviation provided the research field for Orasanu and Fischer's (1997) model of pilot decision making, whereas Klein's model was originally developed on the basis of observations of fireground commanders. However, there is over-riding agreement between the models on certain key elements. These elements include situation assessment, mental imagery, and dynamic decision making.

A simplified framework (see Figure 11.1) illustrates the relationship between situation assessment and decision making strategy. Situation assessment feeds into a continuum of decision making strategies that range (in terms of decreasing cognitive resources/'thinking power' required) from creative, through analytical, then rule-based, to recognition-primed.

These two steps – 'what's the problem', and 'what shall I do', are discussed in turn in the following sections.

Figure 11-1 Relationship between situation assessment and decision making strategy

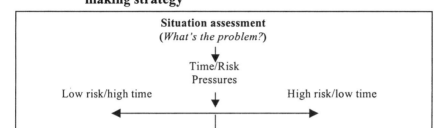

Situation Assessment: Step 1: What's the Problem?

Situation assessment, a key feature of most NDM models, is considered paramount to effective decision making (Cannon-Bowers & Bell, 1997; Endsley & Garland, 2000), where the first step in the decision making task is to evaluate the characteristics of the event correctly. Rapid decisions are made holistically, on the basis of situation recognition and pattern matching through to memory structures. As Endsley (1997) states "In most settings effective decision making largely depends on having a good understanding of the situation at hand" (p. 269). Situation assessment is the process of achieving situation awareness, the product (Adams, Tenney & Pew, 1995), and refers to the acquisition of information, i.e. the integration of cues from the environment, being interpreted on the basis of pre-existing knowledge leading to meaning being given to the cues. It is broader in scope than diagnosis, which typically refers to searching for the cause of abnormal symptoms (Roth, Mumaw, & Lewis, 1994). In the Fire Service, this may be referred to as "size up" (Brunacini, 1985). See also Lodge's chapter (this volume) where he discusses how pilots have to maintain situation awareness.

Endsley (1997) proposes that situation awareness comprises three different levels:

Level 1 consists of perceiving the status, attributes, and dynamics of relevant elements in the environment. For any individual, this relates to their awareness of their own situation and that of others. For example, a pilot needs to be aware of other aircraft, mountains, or warning lights.

Level 2 combines and goes beyond the elements from Level 1. This level relates to comprehending the relevant cues from the environment and their significance in terms of the individual's own goals (including activities such as diagnosis or fault identification). Again, using an aviation analogy, a military pilot has to understand what the movements of enemy aircraft in a certain pattern at a certain location means about their objectives.

Level 3 further combines Levels 1 and 2, and refers to projecting the future activity of the elements in the environment. The offensive actions of enemy aircraft (based on formation and location) allows a pilot to make a projection about likely attack. Many of the commanders in the earlier chapters describe this level of anticipation in their thinking. Larken has described it neatly as 'what-ifing' or 'intuitive extrapolation'.

Decision makers may take the correct decision based on their perception of the situation, but this perception may be incorrect. This is a very different problem from when the situation is understood correctly and a poor choice is made for the best possible action. Thus the role of situation awareness is particularly crucial to command decision making as, in times of high mental workload and stress, it appears that any 'loss' of situation awareness may negatively influence the decision making process. Faulty or inadequate situation assessment has been proposed to have contributed to errors by individuals in previous disasters, rather than choice of action

(Rouse, Cannon-Bowers, & Salas, 1992). Frequently cited examples of accidents in complex systems include the USS Vincennes (Fogarty, 1988), the Three Mile Island accident (Rubinstein, 1979), the Pan Am and KLM collision at Tenerife (Roitsch, Babcock, & Edmunds, 1978), the explosion of the process plant at Flixborough (Parker, Pope, Davidson, & Simpson, 1974), and the crash of Pan Am Flight 401 in Miami (National Transportation Safety Board, 1973).

Situation assessment is also compatible with one part of Boyd's 'OODA loop' used in the US Marine Corps to describe the decision making process (Paradis, Treurniet, & Roy, 1998). The OODA loop consists of a cycle of four elements:

- Observe – gather information and data from the situation
- Orient – assess the situation and process data about the current situation
- Decide – make a decision and select the course of action to execute
- Act – implement the selected course of action.

The OODA loop is smaller in quick tempo periods, as all elements of the loop occur rapidly, but is less constricted in slower tempo periods.

The use of processes such as the OODA loop assists on-scene commanders in effective incident command, and highlights the importance of time and risk assessment, as discussed in further detail below.

Assessment of Time and Risk

As Orasanu and Fischer (1997) pointed out, as part of situation assessment, factors such as level of risk and time available for making decisions must also be taken into consideration. These situational aspects were found to influence decisions made and actions taken by expert pilots, as time and risk called for an immediate response whether or not the problem was fully understood. When risk and time pressures were low, further diagnosis could take place and various options could be considered. Pilot behaviour in simulators was also concurrently examined to determine what constitutes effective flight crew decision making, and specifically, what conditions pose problems and lead to poor decisions. Their observations confirmed that the more effective crews tailored their decision strategies to the situation. Moreover, effective crews also employed generic strategies beneficial in all decision contexts, such as good situation assessment, contingency planning, and task management. Use of such strategies allowed decision makers to 'buy' more time in which to make a good decision, whereas lower performing crews were driven by time pressures and situational demands rather than managing their 'windows of opportunity'.

In the UK dynamic assessment of risk and operational risk management is encouraged through techniques such as the Dynamic Assessment Flowchart and Dynamic Assessment Method (Home Office, 1998), as shown in Figure 11.2. The Fire Service Manual (Home Office, 1999) describes and details the concept of dynamic risk assessment, and contains examples of an operational risk assessment process, as developed by Arbuthnot in the West Yorkshire Fire Service. The Manual also presents a description of the roles of the other emergency services, an introduction to the psychology of command, legal considerations in command and control, as well as examples of the application of the incident command system.

The aim of the development of a model of dynamic risk assessment is to minimise the risk to firefighters and to provide an acceptable level of protection at operational incidents. Incident commanders are required to:

- Evaluate the tasks and persons at risk;
- Select systems of work;
- Assess the chosen systems of work; and to
- Re-assess systems of work and additional control measures.

The purpose of dynamic risk assessment is to assist the decision maker to weigh up whether risks are proportional to the benefits. Tasks can be undertaken once it has been ensured that goals, both individual and team, are understood, responsibilities have been clearly allocated, and safety measures and procedures are understood.

Figure 11.2 Dynamic assessment flowchart

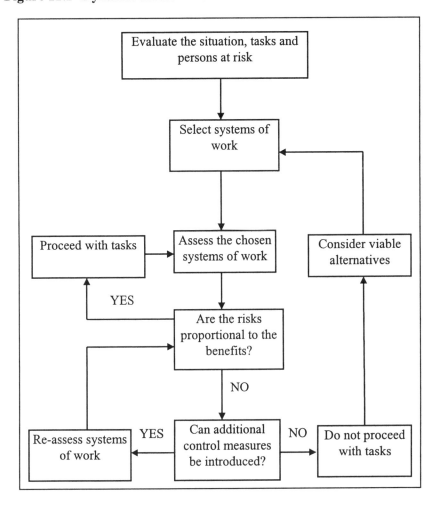

Source: A Fire Service Guide, 1998; p. 13, HMSO. Crown copyright is reproduced with the permission of the Controller of Her Majesty's Stationery Office.

Situation awareness is the crucial initial stage of effective command decision making. Understanding the situation, as well as weighing up available time and possible risks, inextricably impacts on the course of action selected. Commanders, with good situation awareness "will have a greater likelihood of making appropriate decisions and performing well in dynamic systems" (Endsley, 1995, p. 61).

Four Decision Making Strategies: Step 2: What Shall I do?

Once an initial situation assessment has been reached, the next step for the commander is to decide on a course of action (CoA) appropriate to the identified situation. The choice of decision making strategy being dependent upon the influencing factors at that particular phase of the incident, such as demands of the situation, implementation of response procedures, and type, or phase, of the incident (Martin, Flin, & Skriver, 1997). The incident commander requires the ability to adapt behaviour depending on task demands, time pressure, and information constraints (Klein, 1989; Serfaty & Michel, 1990). As Figure 11.1 shows, a CoA can be selected by using different decision processes. Based on studies from aviation (Orasanu, 1997), the military (Pascual & Henderson, 1997; Schmitt, 1994), and the police (Crego & Spinks, 1997), it appears that incident commanders may use one of four decision strategies, depending on their assessment of the available time and the level of risk:

- Recognition-primed (intuition/ gut feel) (If X then Y – little conscious effort needed to retrieve Y)
- Procedures (written or memorised) (If X then Y – conscious search)
- Analytical comparison of different CoA options (If X, which Y?)
- Creative (designing a novel CoA) (If X, have no Y, design new Y).

These four decision strategies are ordered in Figure 11.1 in relation to their apparently increasing requirement for cognitive resources. That is, requiring increasing levels of mental concentration, not just to retrieve information from the memory stores (what psychologists call 'Long Term Memory), but to consciously operate on or think about the information retrieved (this is using 'Working Memory'). Intuitive or Recognition–primed decision making is very rapid and seems to require little conscious effort, placing few demands on Working Memory (see Stokes, 1997, p186). (Fallesen (2000, p. 191) calls this "nonconscious use of rules or knowledge".) In contrast, analytical decision making that involves considering the pros and cons of several options simultaneously, or creative decision making that designs a novel course of action, both require extensive cognitive resources and have a heavy demand on conscious processing (Working Memory).

The decision strategies selected are likely to change during the event, and more than one strategy may be being employed over a given time period, as the event goes through periods of 'crisis phasing' (Crego & Spinks, 1997), where there are periods of severe time constraint juxtaposed with periods of relative quiet and calm. This is a cyclical process with ongoing attention to monitoring, maintaining, and sharing situation awareness. These four decision processes are discussed below in turn.

Recognition-Primed Decision Making

The Recognition-Primed Decision making (RPD) model (Klein, 1998) prototypical of NDM models, was initially developed from studies into decision making by urban fireground commanders (Klein, Calderwood, & Clinton-Cirocco, 1986). As the US Army were sponsoring the research, this particular decision making role was considered to be representative of experienced incident commanders in operational settings where decisions often need to be made quickly in complex, dynamic and hazardous situations, with limited time and information. During interviews, the incident commanders stated that they had tended to concentrate on assessing and classifying the situation. Once they recognised that they were dealing with a particular type of event, they usually also knew the typical response to tackle it. The feasibility of that course of action would then be evaluated, by imagining how they would implement it, to check whether anything important might go wrong. If they envisaged any problems, then the plan might be modified but only if they rejected it, would they consider another strategy.

Recognition-primed, or intuitive, decision making is a fast decision making strategy, which occurs when there may not be an actual written rule or procedure. The incident commander rapidly recognises the type of situation and immediately recalls an appropriate course of action, based on prior experience (Kerr, 1994). (See Claxton, 1997, for a broader discussion of intuition).

On the basis of these findings, Klein et al., (1986) proposed that decisions were based on previous knowledge and expertise that allowed experienced decision makers to recognise a situation as typical and recall the appropriate response to deal with it. Assessing and classifying the situation was emphasised rather than generating options. With experience, decision makers could usually read the situation so that the selection of a course of action was obvious. The RPD model has three levels: the simplest level relates to a condition/action sequence where a decision maker sizes up a situation and responds with the initial option identified. The second level refers to an unclear situation: the skilled decision maker then relies on a story-building strategy to mentally simulate the events leading up to the situation. Finally, the third level describes how decision makers can evaluate a course of action without comparing it to others and where the course of action is mentally simulated – to see if it will work. Only if it appears to have unacceptable consequences is another option considered.

Subsequently, the same pattern of results supporting the use of RPD has been found in other domains where decision makers are faced with similar demand characteristics (e.g. tank platoon captains, naval warfare commanders, intensive care nurses) (Klein, 1998), pilots (Stokes, 1997), army officers (Fallesen, 2000), fire service officers (Burke & Hendry, 1997; Tissington, Flin, & McGeorge, 1998), and police officers (Crego & Spinks, 1997).

Procedure-based Decisions (e.g Standard Operating Procedures, SOPs)

Another decision strategy is the use of procedures or rules, such as Standard Operating Procedures (SOPs) or Emergency Operating Procedures (EOPs). These methods involve the identification of the problem faced and the retrieval from memory or published manuals/checklists of the rule or taught method for dealing with this particular situation. These are widely used in aviation and in industry and are frequently practised in training, in order that the critical rules may be committed to memory. In high reliability industries SOPs are often an integral part of installation design and are devised against specific acceptability criteria. SOPs are generally well known, and can be recalled easily from memory, or if less familiar can be supported by the use of check sheets, manuals, or cue cards. Rules and procedures are in many cases, an excellent method of decision making in emergencies as they have normally been written by experts on the basis of careful analysis of alternative courses of action. Their application requires identification of the problem (not always straightforward, as discussed above), then retrieval of the matching action. This requires more cognitive processing than RPD, but much less than analysis of alternative options or creating a new option. Hence the heavy reliance on procedures in time pressured (e.g aviation) and safety-critical domains (e.g nuclear industries), although both these domains acknowledge that unforeseen situations or unusual combinations of events can arise.

Merely having SOPs in place does not necessarily guarantee compliance (Dobson, 1995). As Reason (1990) points out, human beings can be susceptible to performing 'violations', i.e. deliberate, but not necessarily reprehensible, deviations from practices deemed necessary to maintain the safe operation of a potentially hazardous system such as operating procedures, codes of practice and rules. This is particularly likely to happen if the rules are out of date or if they have been superceded by custom and practice. SOPs for emergency management should be regularly updated and should be designed in such a way that the crucial importance of being able to improvise is still recognised, and the SOP provides a suitable framework rather than repressing flexibility (Dror, 1988). For example, an investigation of decision making in simulated emergencies by Offshore Installation Managers (OIMs) in the North Sea oil industry (Skriver & Flin, 1997), found that their decisions appeared to be made on the basis of using memorised SOPs primarily as guidelines, rather than as fixed rules.

A trend exists in many high risk settings to create SOPs for every predictable event or crisis. However, if a commander is used to relying on finding a prescribed procedure in a manual for every situation, then a problem may emerge when a novel emergency is encountered. Almost by definition, a crisis arises because the events are unanticipated and no simple procedures exist to remedy the situation. Moreover, an over-reliance on SOPs may lead to the situation where they are viewed as hard and fast rules that must be followed blindly, and the exploration of alternative ideas may be discouraged. As Skriver and Flin (1997) conclude, increasing the number and specificity of emergency SOPs in industrial settings, may actually weaken managers' ability to deal with the unexpected crisis.

In domains where there are few prescribed SOPs, then a more flexible approach may be warranted. Rather than depending on finding answers in a rule-book during an emergency, situational cues may be matched to existing patterns stored in memory, with the SOPs used as guidelines. This, of course, requires a good knowledge of procedures combined with experience, which will normally be developed through training, e.g. exercises, simulator training, and involvement in safety management systems. While standard procedures are extremely helpful, it is important to ensure that decision making in emergency situations is based on sound foundations, not simply on blind application of rules.

Analytical/ Option Comparison

Analytical decision making typically occurs when neither time nor risk are limited, and where information to evaluate alternative options is available. Decision making using this process involves a process of rigorously evaluating, then selecting among possible actions (Gilhooly, 1988). Examples of analytical decision making include choosing a holiday destination or a new car; or policy making in health, economics, law, and environmental issues (see Connolly et al, 2000, for a review). Analytical decision making requires a full situation assessment and rigorous information search, and comparison of several courses of action retrieved from memory, thus resulting in a high load on working memory. This therefore takes longer to accomplish and is generally only feasible in times of relative quiet with minimal distraction (and preferably with paper and pencil to hand). Heuristics (rules of thumb or simplifications) tend to be used to reduce the memory load as various options are weighed up. This is usually the optimal method of reaching a decision, as all alternatives are considered and carefully evaluated. There are innumerable theories and complex mathematical methods to aid in the perfection of this process, ranging from Bayesian statistics to Multi-attribute utility theory. Of necessity, analytical decision making is time consuming, requires high cognitive resources and is extremely difficult to do in noisy, dynamic environments. Therefore it tends to be utilised by commanders in phases of lower time pressure (e.g. police hostage negotiations), or in planning phases of emergency preparedness, or at the strategic level of command where less immediate, longer term plans are devised.

Creative Decision Making

A fourth category of decision making is 'creative decision making'. This is the process used in instances where the decision maker not only has to diagnose an unfamiliar situation, but also has to design a novel course of action, as no stored rules or memories of suitable actions are available. This is similar to Orasanu's (1993) concept of 'creative problem-solving for ill-defined problems', and Svenson's (1997) 'Level 4 decisions' where decision alternatives have to be elicited or created in response to a new and unfamiliar problem. Examples of creative decision making in non-routine emergencies, involving the generation of previously un-thought-of courses of action, are cited as the hydraulic systems loss by United Airlines Flight 232 (Orasanu, 1993), and the oxygen tank explosion aboard Apollo 13 (Klein, 1998).

To date, very little research has been conducted into this type of decision making in emergencies. It is anticipated that creative decision making would require considerable cognitive resources to work out a new solution from first principles. (This is very different from recognising that the situation is similar to something encountered previously, which is analogical reasoning and essentially the basis of a recognition-primed decision.) Creative decision making is unlikely to be attempted in an emergency unless the commander has no other solutions that will satisfactorily contain the situation and has an hour or more to think about the problem. Creative decision making would appear to be very rarely used by incident commanders working under time pressure, and is in fact what should be performed during the planning and procedure development phases of operations management when there is adequate time to examine, test and evaluate novel courses of action (Kersholt, 1997).

A Decision Taxonomy for Incident Command

Decision making at the three levels of the incident command structure (strategic/tactical/operational) (see Arbuthnot, this volume, for details) can be differentiated in terms of the balance of decision making strategies identified above (Home Office, 1999).

Strategic command Decision making at the strategic management level would appear to be less time constrained, less immediately critical, more analytical, and possibly more consultative, albeit still pressured. The decision making style may either be more analytical as time is available in which to consider various options, or be more creative as the situation may feature a number of novel elements or developments not previously encountered by the strategic commander. The aim is to devise the optimal solution for the situation, taking into account the wider and longer-term implications.

Tactical command At this level, decision making will generally be more immediate, less analytical, but with heightened time and risk pressures. Decision making will probably be based more on condition-action matching, or rule-based reasoning (i.e. If ... (condition)...Then... (action)...). This style is characterised by controlled actions derived from procedures stored in memory. The onus for tactical decision making is to maintain a good mental model of the evolving plan and unfolding events. The tactical commander may need to 'create' time to engage in reflective thinking, and when necessary, use more analytic decision strategies to evaluate alternative courses of action.

Operational command Decision making at the operational level, or "sharp end", is, often out of necessity, faster, more critical, and time and risk-pressured. Decision making takes place in real time, and operational commanders may have to react rapidly to situational demands. Decision making therefore tends to contain rule-based as well as intuitive elements. Under time and risk pressure, decisions may primarily be made on the basis of pattern recognition (e.g. RPD) of the situations encountered. Moreover, situation awareness is critical, as performance depends on rapid identification of the situation and fast access to stored patterns of pre-programmed responses. When time pressures or risk are low, a more analytical decision making strategy may be utilised.

Fredholm (1997) comments that, at major incidents for the fire service, much of the on-scene commander's decision making is reactive, in that it is directed towards the immediately visible problems, especially limited situation problems (see Table 11.1).

Table 11.1 Categories for tactical problem alternatives

Situation	Description
4	Unlimited situation. Weak resources. Great demands on tactical judgement.
3	Limited situation. Weak resources. Less demanding tactically.
2	Limited situation. Critical resources. Great demands on tactical judgement.
1	Limited situation. Strong resources. Less demanding tactically.

Source: Fredholm, 1997, Figure 1, p110 Reprinted with permission of Ashgate.

He distinguishes between limited and unlimited situations on the basis of whether the situation can be surveyed directly or not (i.e. immediately visible), and the resources immediately available. This results in four categories of situational demand on the decision maker. Limited situations call for rapid decision making. However, in a situation where time and space are unlimited, anticipation, abstraction, and visualisation of the development of the situation are actually required, and a more analytical, long-term decision making strategy is more appropriate. Thus, where time, risk, and quality of information allow, decision makers may be able to select from a number of generated decision options and pick the most suitable (Crego & Spinks, 1997 make a similar point in relation to police officers).

Choice of a particular decision making strategy is influenced by factors such as amount of time and level of risk, availability of known rules or procedures, availability of information, and complexity of, and familiarity with, the situation. For the incident commander, pre-planning, experience, and training help to build up a repertoire of well-rehearsed CoAs, which can readily be implemented. Of course, completely novel situations can and do arise. The ability to assess the situation as fully as possible, and to take the best advantage from the time available, allows the incident commander to exercise control and to effectively command such situations.

A similar taxonomy for military command decision making has recently been proposed by Fallesen (2000). Based on a series of research studies of US army commanders' decision making, he argued that a stronger distinction should be made between different modes of thinking that underlie decision making. Fallesen distinguishes between formal and informal approaches to thinking. Formal thinking is more structured, proceduralised, precise, and tends to be limited to well-understood problems. Doctrinal training, i.e. rules and knowledge, and procedural-regulated training fall into this category. Informal approaches, on the other hand, are more descriptive than prescriptive, and illustrate that thinking is not always deliberate, conscious or rational. Informal thinking is influenced by the individual's beliefs about knowledge, attitudes, and skills, and is suited to dynamic and complex everyday problems. The use of recognition, or non-conscious use of rules or knowledge, and dominance, based on Montgomery's (1993) 'search for dominance' strategy, which can involve analysing, testing or elaborating a hunch or intuition, are emphasised in informal thinking approaches. Fallesen's framework shares similarities with the taxonomy proposed in Figure 11.1, with the same analytical, procedural, and recognitional components, but he replaces the postulated creative option with dominance.

Fallesen's taxonomy of military problem solving approaches acknowledges that battle command, or indeed any incident command, cannot be entirely managed by formal problem solving approaches, but that there is a need for informal methods, characterised by flexibility, adaptability, and a knowledge of principles rather than procedures. In addition to the US army's traditional reliance on formal doctrines and decision procedures for effective command and control, Fallesen asserts that training in conceptual thinking is also required to promote versatility, speed, and efficiency of decision making.

Stress and Decision Making

Problem solving and decision making in demanding real-world situations can be susceptible to acute stress effects which manifest in a variety of ways depending on the type of decision (Svenson, 1997). The negative effects of an overload of acute stress include attentional tunnelling, working memory loss, and restrictions in long term memory retrieval, with simple strategies being favoured over more complicated ones (Ben Zur & Breznitz, 1981; Wickens, 1996; Zakay & Wooler, 1984). "The underlying assumption is that stress can lead to errors, poor performance and bad decisions." (Orasanu & Fischer 1997, p.43). For example, Svenson (1997) suggests that stress may invoke a re-allocation of "energetic resources from the decision process to stress coping processes..." and ... "may change the goals in a decision situation and the structuring and processing of information." (p. 308). However, acute stress does not necessarily always have a detrimental effect on decision making, rather stress may affect the way information is processed. Klein (1998) argues that some of those changes in strategy in response to stress are in fact adaptive. They reduce and select the information being attended to and processed, in response to high time pressure and reduced cognitive capacity.

The Yerkes-Dodson Law illustrates the stress/performance relationship as an inverted 'U' shape, where performance initially rises with increasing pressure, but then declines as pressure continues to increase and the individual begins to experience stress (Fisher, 1986). Stokes and Kite (1994) question the almost uncritical acceptance of the "ubiquitous 'U'" (p. 32). They comment that, as the Yerkes-Dodson curve originally relates to arousal, which is a separate phenomenon from stress, stress-related changes in cognitive performance depend on an interaction of the demands of the task, the stress characteristics of the situation, and the individual's subjective stress responses. Certainly a commander's experience of stress will depend on a number of individual factors, such as personality, training and experience (see Flin, 1996 for a chapter on the causes, effects and buffers of stress for incident commanders).

Where acute stress is likely to have a particularly detrimental effect on decision making is in the use of analytical or creative strategies, as these generally require extensive cognitive effort, especially Working Memory resources which are significantly depleted under stress as attentional capacity diverts to monitoring the threat (Stokes, 1997). Recognition primed or rule-based decision strategies, in contrast, may be more resistant to the effects of acute stress, as these require less cognitive effort (Klein, 1998). The postulated effects of stress on decision making strategies are illustrated in Figure 11.3 below.

Figure 11.3 Relative effects of stress on decision making strategy

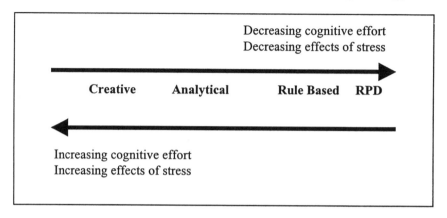

More research is needed in this area, particularly in relation to identifying prime causes of stress (stressors) for commanders and in developing techniques to enhance their stress-resistance. (See Breakwell, 2000, for an interesting account of the causes of stress experienced by senior British Army commanders in the Gulf, Bosnia and Northern Ireland).

Training for Command Decision Making

The major distinction between members of the emergency services and of industrial ERO personnel is that of probability versus possibility. Whereas personnel in the emergency services, continuously experience and train for incident response, ERO personnel, due to the infrequent occurrence of industrial emergencies, rely on regular but occasional training interventions to prepare for effective emergency response. However, the premise for incident command training is similar in both instances. The longer term benefits for the individual include encouragement to use initiative, to take personal responsibility and to bring together and reinforce training and actual experience. For the organisation, the benefits of command decision making include identifying areas for further review, supporting personnel in the command process, and improving the knowledge of those who may be required to take on the role of incident commander or provide command support at incidents. (An organisation which could experience an emergency and does not train or assess the competence of its incident commanders would be extremely exposed in the case of a Public Inquiry or other consequent litigation).

To effectively manage an incident, the incident commander must be able to (i) assess the situation, (ii) decide on a course of action, (iii) implement that plan of action in line with operational constraints and (iv) monitor the operation thereafter. In order to be effective, training in incident command skills, according to McLennan, Pavlou and P. Klein (1999), involves four elements:

- Provision of a simple, robust conceptual scheme of incident command
- Opportunity to actively practise incident command in a setting that adequately simulates the psychological demands on the commander
- Provision of feedback about the effectiveness of command and control decisions and actions
- Opportunity for guided reflection and self-appraisal.

Psychologists working with the US military point out that the emphasis for cognitive skills development is shifting from "what to think" to "how to think" (Noble, Fallesen, Halpin, & Shanteau, 2000). Their aim being to assist battle commanders to cope in an environment that is increasingly complex (e.g. flow of information, joint forces, multiple missions, increased interaction with outside agencies, and media presence) by being more adaptive, flexible, and versatile when making decisions. The US Army Research Institute for the Behavioral and Social Sciences (ARI) comment that battle command thinking, and subsequent command performance, can be improved through a program of instruction on 'practical thinking' (Fallesen, Michel, Lussier, & Pounds, 1996). The practical thinking course intentionally avoids prescribing procedures for thinking, but instead introduces concepts and techniques for performing battle command. The aim is to encourage reflection, flexibility, discovery learning, critical and creative thinking.

Incident command training usually comprises on-the-job training, such as exercises, drills or observing (or shadowing) existing incident commanders, often supplemented with training courses at specialist centres or colleges. An integral and vital part of any training intervention, including training for incident command and emergency management, is that of augmented and constructive feedback. However, feedback presented at the end of an emergency exercise, tends to focus on the outcome rather than the process of performance. Although feedback presented throughout an exercise can be effective, care must be taken to ensure that it does not interrupt or interfere with performance. Nevertheless, when errors are pointed out, the trainee should be able to infer the lesson to be learned from the error (Cannon-Bowers & Salas, 1998). However, feedback does not necessarily lead to acquisition of knowledge, and the trainee must be psychologically open to, receptive of, and reflective about the feedback message in order to alter performance (McLennan et al., 1999).

A relatively novel system to provide feedback on fireground command performance during an incident, is that of head-mounted video. McLennan et al (1999) have developed an approach to training based on video footage, giving the field of view from the incident commander's helmet, during an emergency exercise. Video-cued replay is used for debriefing and feedback purposes. This provides a powerful cue for commanders to recall the basis of their incident control decision making, identifying uncertainties, self-questioning, and self-doubt. This in turn was considered as a positive method by which to receive feedback suggestions about improvements in performance.

Flin (1996) provides a comprehensive review of training methods for incident command, ranging from background reading, lectures, to computer-based training, simulators, and full-scale exercises. However, training programmes are useful for imparting the rules and procedures required for skills and knowledge, but it is not necessarily the case that trainees are taught to make better judgements or decisions (Klein, McCloskey, Pliske, & Schmitt, 1997). In addition, being taught to adhere to and apply operating procedures may not provide the opportunity to explore alternative ideas that may be required when dealing with a novel emergency (Skriver & Flin, 1996). A need exists for training interventions specifically directed towards enhancing individual and team skills, leading to improved efficiency and safety, reduced error, and enhanced overall task performance (see Salas et al, this volume). Training is particularly required for effective decision making in emergencies based on situations with unexpected elements and unusual combinations of problems. To be effective, this training is best directed towards increasing expertise by supporting a decision maker's existing strategies rather than teaching new more formal strategies (Klein, 1997b).

Shortcomings in training methods arise when, for example, either too much effort is expended on the replicating the physical aspects of the incident ground to the detriment of the training objectives, or the training course content is too specific, theoretical or complex. Case studies, when they do exist (see this volume and McCann & Pigeau, 2000), and background reading are very useful training materials but obviously have limitations in conveying the personal psychological demands of incident command. Full-scale emergency exercises, on the other hand, offer excellent opportunities to practise incident command training, but they can be costly in terms of time, resources and logistics. In many industrial locations, (e.g. flight decks, control rooms on power plants) it is not possible to simulate fully emergency conditions and full fidelity simulators have to be developed. These are excellent for incident command training but are extremely expensive.

Computer-based simulations offer a degree of realism without the associated costs, for example, the Vector Command (VECTOR Command UK, 1999) as used by a number of UK Fire Brigades, and command team systems such as Minerva and Hydra (see Crego, this volume; Lines, 1999) as used by the London Fire Brigade and the London Metropolitan Police. These are being used increasingly by the UK emergency services to provide trainees with practice in the decision skills inherent in incident command and management.

A relatively new training intervention for decision making in emergencies, is that of Tactical Decision Games (TDGs) (Crichton, Rattray, & Flin, 2000b). These 'games' have traditionally been used in military settings (Schmitt, 1994; Schmitt & Klein, 1996) but have more recently been adapted and developed for non-military use.

Tactical Decision Games A Tactical Decision Game (TDG) is a low-fidelity facilitated simulation, based on scenarios ranging in complexity and technicalities, of incidents that may occur during emergency response and are designed to exercise relevant non-technical skills, primarily decision making. Participants in the TDG, usually four to ten individuals, are presented with brief written details of a scenario by a Facilitator and may take on assigned roles. The scenario, in which information can be misleading, ambiguous, or missing, culminates in a dilemma that must be resolved. Participants must decide upon actions to be taken to manage the situation. As only a limited amount of time (e.g. two to five minutes) is available for each decision – they must be made quickly while under some degree of pressure. Each participant generates their own solution to the dilemma, then differences and similarities between solutions are discussed. After the initial scenario presentation, and during the discussion, the Facilitator can introduce additional information or contingencies, in an incremental manner, that will increase the complexity and uncertainty of the situation. The objectives of TDGs can be summarised as:

- To exercise and practise decision making skills and illustrate key operating principles.
- To boost expertise in decision making and judgement.
- To assist participants to develop a shared understanding and recognition of possible problems.
- To build up a repertoire of patterns which can be quickly recognised and acted upon, particularly during emergency situations.
- To practise non-technical skills such as decision making, communication, situation awareness, stress management, and teamwork (Klein, 1998; Klein & Wolf, 1995; Schmitt & Klein, 1996).

A prevailing principle of TDGs, however, is for all participants to develop a shared understanding and recognition of possible problems for emergency management. For participants to learn from their experience during the TDG scenario, it is essential that they analyse and evaluate what happened, going beyond simply naming the strong and weak points of their own performance. Powerful insight can be gained by analysing *why* decisions were made or actions taken, including factors that either enabled or hindered their success. The Facilitator has at his/her disposal learning tools, (e.g. those developed by Pliske, McCloskey & Klein, 1998) to fully explore the decision making, and other non-technical skills, which emerged during the session. These learning tools consist of a technique for identifying decision requirements (Decision Requirements Exercise), a method for reflecting on the decision making in training events (Cognitive Critique), a method for mentally simulating plans (Post-mortem and/or Pre-mortem Exercise), and a method for leaders to obtain feedback on the expression of intent (Commander's Intent).

Crew Resource Management A more general training programme suitable for incident commanders is Crew Resource Management (CRM). It is taught in the classroom and practised in simulators, and it focuses on decision making and situation awareness, as well as communication, leadership, team working and stress management. This human factors training was initially developed by the aviation industry (Helmreich & Foushee, 1993), and has now been adopted by a number of other high reliability organizations, including the offshore oil industry, the nuclear power industry, air traffic control, surgical medicine and the merchant navy (Salas, Bowers & Edens, 2001). Although CRM was originally designed to improve aviation safety, its basic tenet of enhancing interpersonal and cognitive skills can also be applied in other situations, such as emergency response where dependence on team activities is high.

The core CRM skills can also be called non-technical skills. Research in domains such as aviation and medicine has shown that where these skills are deficient, this is a prime cause of accidents and adverse outcomes (see Lodge, this volume; Helmreich & Merritt, 1998). Current research projects are developing methods of training and evaluating these non-technical skills using behavioural marker systems (Flin & Martin, 2001; Salas, et al 2001). A taxonomy called NOTECHS (Table 11.2) has been developed for European pilots' non-technical skills in multi-crew aircraft(O'Connor et al, in press; van Avermaete and Kruijsen 1998). The system consists of four categories, two of social skills (co-operation; leadership/ management skills) and two of cognitive skills (situation awareness; decision making). Each category is subdivided into smaller skill elements. This system is designed for the proceduralised domain of the flightdeck, but the main skill categories are applicable to emergency command (outlined earlier) and its framework may offer a useful basis for training and assessing the core non-technical skills for command.

Table 11.2 The NOTECHS framework (van Avermaete et al, 1998)

Categories	Elements
Co-operation	Team building and maintaining Considering others Supporting others Conflict solving
Leadership and managerial skills	Use of authority Maintaining standards Planning and co-ordinating Workload management
Situation awareness	System awareness Environmental awareness Anticipation
Decision making	Problem definition / diagnosis Option generation Risk assessment Outcome review

In summary, as actual experience of incident command is rare for many who may be called upon to command a demanding incident, the onus for training providers is to ensure that training in incident command is focussed, effective and edifying. Although the content of training programmes may vary across organisations, the ultimate aim is to ensure that the incident commander is proficient and capable, with the ability to assess the situation, to implement procedures for dealing with an emergency, and to make command decisions. Training should therefore provide not only a knowledge of emergency procedures but should also imbue the incident commander with the competence and confidence to take command and to control in the situation.

Conclusion

Incident command not only relies upon the implementation of a prepared emergency plan and procedures, but also depends upon the skills of the incident commander and command unit team members. The objective of this chapter has been to present, from the basis of current psychological research, the decision making skills required by incident commanders.

The aim for those tasked with command decision making is, somehow, to make the 'best' decision possible and there is growing recognition that command decision making is about making satisfactory rather than perfect or optimal decisions. As command decision making is often time pressured, in difficult, distracting conditions, with inadequate information, then flexible and versatile decision making is essential to ensure effective command performance. This necessitates the use of decision making strategies ranging from intuition-based through rule-based, to analytical or creative, dependent upon situational factors such as time and risk.

At all levels, effective command decision making is no mean feat. Yet, as can be seen from the personal experiences of contributors described earlier in this book, it is a requirement demanded of many incident commanders, both civilian and military, across a variety of demanding events. The acquisition of expertise, not only through experience but also through directed and focussed training, can only serve to improve the quality of command decision making performance. On-going psychological research continues to expand our understanding of the skills required by this fascinating and all-too necessary topic of incident command. The findings of this research are being integrated into application areas such as the development of training tools and interventions, support, and selection of incident commanders.

Key points

- Effective incident command depends upon the skills, abilities, and competence of both the incident commander and the command unit team, as well as the establishment of an incident command structure.
- The strategic/tactical/operational command structure generally invokes the use of different decision making strategies, particularly in terms of time available to make decisions. The strategies can be visualised on a continuum ranging from creative and more analytical (strategic command) through procedural and rule-based (tactical command) to a more intuitive decision making strategy (operational command).
- Effective command decision making creates demands on associated skills such as situation awareness as well as time and risk assessment. In addition, incident commanders must be able to cope in an

increasingly complex environment by becoming more adaptive, flexible, and versatile when making decisions.

- Training for incident command requires four key elements, namely: the provision of a simple, robust conceptual scheme of incident command; the opportunity to actively practise incident command in a setting that adequately simulates the psychological demands on the commander; the provision of feedback about the effectiveness of command and control decisions and actions; and the opportunity for guided reflection and self-appraisal

- Training interventions for command decision making range from high-fidelity techniques, such as full fidelity simulators, to low-fidelity methods, such as Tactical Decision Games (TDGs). The aim of all such training interventions is to practise and exercise decision making and associated non-technical skills, leading to more effective incident command.

Acknowledgement: We are most grateful for the constructive comments on earlier versions of this chapter received from Kevin Arbuthnot, Jeremy Larken, Gary Klein and Eduardo Salas.

Websites for information on command decision research and training:

Industrial Psychology Group, University of Aberdeen
www.psyc.abdn.ac.uk/serv02.htm

Klein Associates
www.decisionmaking.com

MINERVA/HYDRA (computer-based command team simulator)
www.essenet.demon.co.uk

VECTOR (computer-based command training simulator)
www.vectorcommand.co.uk

References

Adams, M., Tenney, Y., & Pew, R. (1995) Situation awareness and the cognitive management of complex-systems. *Human Factors*, 37(1), 85–104.

Ben Zur, H., & Breznitz, S. (1981). The effect of time pressure on risky choice behaviour. *Acta Psychologica*, 47, 89–104.

Breakwell, G. (2000) In C. McCann & R. Pigeau (Eds.), *The human in command. Exploring the modern military experience*. New York: Kluwer Academic/Plenum Publishers.

Brunacini, A. (1985). *Fire command*. Quincy, MA: National Fire Protection Association.

Burke, E., & Hendry, C. (1997). Decision making on the London incident ground. An exploratory study. *Journal of Managerial Psychology*, 12, 40–47.

Cannon-Bowers, J. & Bell, H. (1997). Training decision makers for complex environments: Implications of the naturalistic decision making perspective. In C. Zsambok & G. Klein (Eds.), *Naturalistic decision making*. Mahwah, NJ: Lawrence Erlbaum Associates.

Cannon-Bowers, J. & Salas, E. (Eds.). (1998). *Making decisions under stress. Implications for individual and team training*. Washington, DC: American Psychological Association.

Channel Tunnel Safety Authority. (1997). *Inquiry into the fire on heavy goods vehicle shuttle 7539 on 18 November 1996*. London: CTSA.

Claxton, G. (1997). *Hare brain. Tortoise mind. Why intelligence increases when you think less*. London: Fourth Estate.

Connolly, T., Arkes, J. & Hammond, K. (2000). *Judgment and decision making. An interdisciplinary reader*. Cambridge: Cambridge University Press.

Crego, J., & Spinks, T. (1997). Critical incident management simulation. In R. Flin, E. Salas, M. Strub, & L. Martin (Eds.), *Decision making under stress. Emerging themes and applications*. Aldershot: Ashgate.

Crichton, M., Flin, R. & Rattray, W. (2000a). Training decision makers – Tactical Decision Games. *Journal of Contingencies & Crisis Management, 8*(4), 209–218.

Crichton, M., Rattray, W. & Flin, R. (2000b). Training decision makers – Tactical Decision Games case studies. Paper presented at the Fifth Conference on Naturalistic Decision Making, 26–28 May, Stockholm, Sweden.

Cullen,The Hon Lord. (1990). *The Public Inquiry into the Piper Alpha Disaster* (Volumes I and III (Cm 1310)). London: HMSO.

Dobson, R. (1995). Starting as you mean to go on. A study of compliance with operational procedures during the first 10–15 minutes of incidents (International project report, Brigade Command Course). Moreton-in-Marsh: Fire Service College.

Dror, Y. (1988). Decision making under disaster conditions. In L. Comfort (Ed.), *Managing disaster: Strategies and policy perspectives*. Durham, NC: Duke University Press.

Endsley, M. (1995). Toward a theory of situation awareness in dynamic systems. *Human Factors, 37*, 32–64.

Endsley, M. (1997). The role of situation awareness in naturalistic decision making. In C. Zsambok & G. Klein (Eds.), *Naturalistic decision making*. Mahwah, NJ: Lawrence Erlbaum Associates.

Endsley, M. & Garland, D. (2000). *Situation awareness. Analysis and measurement.* Mahwah, NJ: Lawrence Erlbaum.

Fallesen, J. (2000). Developing practical thinking for battle command. In C. McCann & R. Pigeau (Eds.), *The human in command: Exploring the modern military experience.* New York: Kluwer Academic/Plenum Publishers.

Fallesen, J., Michel, R., Lussier, J. & Pounds, J. (1996). *Practical thinking: Innovation in battle command instruction.* Alexandria, VA: US Army Research Institute.

Fisher, S. (1986). *Stress and strategy.* London: Lawrence Erlbaum Association.

Flin, R. (1996). *Sitting in the hot seat: Leaders and teams for critical incident management.* Chichester: Wiley.

Flin, R. & Martin, L. (2001). Behavioural markers for CRM: A review of current practice. *International Journal of Aviation Psychology, 11,* 95–118.

Flin, R., Salas, E., Strub, M. & Martin, L. (1997). (Eds.) *Decision making under stress: Emerging themes and applications.* Aldershot: Ashgate Publishing.

Fogarty, W. M. (1988). *Formal investigation into the circumstances surrounding the downing of a commercial airliner by the USS Vincennes* (Tech Rep). Washington, DC: Department of Defense.

Fredholm, L. (1997). Decision making patterns in major fire-fighting and rescue operations. In R. Flin, E. Salas, M. Strub, & L. Martin (Eds.), *Decision making under stress: Emerging themes and applications.* Aldershot: Ashgate.

FSAB (1995) *Emergency Fire Services. Supervision and Command N/SVQ Level 3.* Luton: Fire Service Awarding Body.

Gilhooly, K. (1988). *Thinking: Directed, undirected and creative.* London: Academic Press.

Haynes, A. (1992). United 232: Coping with the "one chance-in-a-billion" loss of all flight controls. *Flight Deck, 3* (Spring), 5–21.

Helmreich, R. & Foushee, H. (1993). Why crew resource management? Empirical and theoretical bases of human factors training in aviation. In E. Wiener, B. Kanki, & R. Helmreich (Eds.), *Cockpit resource management.* New York: Academic Press.

Helmreich, R. & Merritt, A. (1998). *Culture at work in aviation and medicine.* Aldershot: Ashgate.

Home Office. (1997). *Dealing with disaster.* (3rd ed.). London: HMSO.

Home Office. (1998). *Dynamic management of risk at operational incidents. A fire service guide* London: HMSO.

Home Office. (1999). *Fire Service Manual.* (Volume 2: Fire Service Operations. Incident Command). London: HMSO.

Kerr, G. (1994). Intuitive decision making at the operational level of command. *The British Army Review, 108,* 5–14.

Kerstholt, J. (1997). Dynamic decision making in non-routine situations. In R. Flin, E. Salas, M. Strub, & L. Martin (Eds.), *Decision making under stress.* Aldershot: Ashgate.

Klein, G. (1989). Recognition-primed decisions. In W. Rouse (Ed.), *Advances in Man-Machine Systems Research.* Greenwich, CT: JAI Press.

Klein, G. (1993). A recognition-primed decision (RPD) model of rapid decision making. In G. Klein, J. Orasanu, R. Calderwood, & C. Zsambok (Eds.), *Decision making in action: Models and methods.* Norwood, NJ: Ablex.

Klein, G. (1997a). The current status of the naturalistic decision making framework. In R. Flin, E. Salas, M. Strub, & L. Martin (Eds.), *Decision making under stress. Emerging themes and applications*. Aldershot: Ashgate.

Klein, G. (1997b). *Making decisions in natural environments*. Alexandria,VA: Research and Advanced Concepts Office, US Army Research Institute for the Behavioral and Social Sciences.

Klein, G. (1998). *Sources of power. How people make decisions*. Cambridge, Mass: MIT Press.

Klein, G., Calderwood, R., & Clinton-Cirocco, A. (1986). Rapid decision making on the fire ground. Paper presented at the Human Factors Society 30th Annual Meeting, San Diego, CA.

Klein, G., McCloskey, M., Pliske, R., & Schmitt, J. (1997). Decision skills training. Paper presented at the 41st Annual Human Factors and Ergonomics Society Conference, Albuquerque, NM.

Klein, G., & Wolf, S. (1995). Decision-centred training. Paper presented at the Human Factors and Ergonomics Society, 39th Annual Meeting, San Diego.

Lines, S. (1999) Information technology multi-media simulator training. *The Police Journal*, LXXII(2), 96–108.

Lipshitz, R., Klein, G., Orasanu, J., & Salas, E. (in press). Taking stock of NDM. *Journal of Behavioural Decision Making*.

London Fire Brigade. (1995). *Training for operational command. A review of simulation methods*. London: Officer Development Group, London Fire Brigade.

Martin, L., Flin, R., & Skriver, J. (1997). Emergency decision making – A wider decision framework? In R. Flin, E. Salas, M. Strub, & L. Martin (Eds.), *Decision making under stress. Emerging themes and applications*. Aldershot: Ashgate.

McCann, C., & Pigeau, R. (Eds.). (2000). *The human in command: Exploring the modern military experience*. New York: Kluwer Academic/Plenum.

McLennan, J., Pavlou, O., & Klein, P. (1999). Using video during training to enhance learning of incident command and control skills. Paper presented at the Fire Service College Conference, 16–17 November, Moreton-in-Marsh, Gloucester.

Montgomery, J. (1993). The search for dominance structure in decision making. In G. Klein, J. Orasanu, R. Calderwood & C. Zsambok (Eds.) *Decision making in action: Models and methods*. Norwood, NJ: Ablex.

National Transportation Safety Board. (1973). *Eastern Air Lines L-1011* (Technical Report N310 EA). Washington, DC: National Transportation Safety Board.

Noble, S., Fallesen, J., Halpin, S. & Shanteau, J. (2000). Exploring cognitive skills of battle commanders. Paper presented at the Fifth Conference on Naturalistic Decision Making, Stockholm, Sweden.

O'Connor, P., Hormann, J.-J., Flin, R., Lodge, M., Goeters, K.-M. & the JARTEL group. (in press). Evaluating pilots' non-technical skills: A European perspective. *International Journal of Aviation Psychology*.

Omodei, M., Wearing, A., & McLennan, J. (2000). Relative efficacy of an open versus a restricted communication structure for command and control decision making: An experimental study. In C. McCann & R. Pigeau (Eds.), *The human in command: Exploring the modern military experience*. New York: Kluwer Academic/Plenum.

OPITO (1997) *OPITO Units of Competence Governing the Mangement of Offshore Installation: 'Controlling Emergencies'*. Montrose: Offshore Petroleum Industry Training Organisation.

Orasanu, J. (1993) Decision making in the cockpit. In Wiener, E.L., Kanki, B.G., & Helmreich, R.L. *Cockpit resource management*. (137–172). San Diego, CA: Academic Press.

Orasanu, J. (1997) Stress and naturalistic decision making: Strengthening the weak links. In R. Flin, E. Salas, M. Strub, & L. Martin (Eds.), *Decision making under stress. Emerging themes and applications*. (43–66). Aldershot, Ashgate.

Orasanu, J., & Fischer, U. (1997). Finding decisions in natural environments: The view from the cockpit. In C. Zsambok & G. Klein (Eds.), *Naturalistic decision making* (343–358). Mahwah, NJ: Lawrence Erlbaum Associates.

Orasanu, J. , Fischer, U., & Tarrel, R. (1993), A taxonomy of decision problems on the flight deck. In R. Jensen (Ed.) *Proceedings of the Seventh International Symposium of Aviation Psychology*, (226–232). Columbus, OH. Ohio State University Press.

Paradis, S., Treurniet, W., & Roy, J. (1998). Environment perception process in maritime command and control. Paper presented at the NATO Symposium on sensor data fusion and integration of the human element, Ottowa, Canada.

Parker, R., Pope, J., Davidson, J. & Simpson, W. (1974). *The Flixborough disaster: Report of the court inquiry* (Technical Report). London: Department of Employment.

Pascual, R., & Henderson, S. (1997). Evidence of naturalistic decision making in command and control. In C. Zsambok & G. Klein (Eds.), *Naturalistic decision making*. Mahwah, NJ: Lawrence Erlbaum Associates.

Pigeau, R., & McCann, C. (2000). Redefining command and control. In C. McCann & R. Pigeau (Eds.), *The human in command. Exploring the modern military experience*. New York: Kluwer Academic/Plenum.

Pliske, R., McCloskey, M., & Klein, G. (1998). Facilitating learning from experience: An innovative approach to decision skills training. Paper presented at the Fourth Conference on Naturalistic Decision Making, Warrenton, Virginia.

Pruitt, J., Cannon-Bowers, J. & Salas, E. (1997). In search of naturalistic decisions. In R. Flin, E. Salas, M. Strub, & L. Martin (Eds.), *Decision making under stress. Emerging themes and applications*. Aldershot: Ashgate.

Rasmussen, J. (1983). Skills, rules and knowledge: Signals, signs and symbols, and other distinctions in human performance models. *IEEE Transactions on Systems, Man, and Cybernetics, 13*, 257–266.

Reason, J. (1990). *Human error*. Cambridge, MA: Cambridge University Press.

Roitsch, P., Babcock, G. & Edmunds, W. (1978). *Human factors report on the Tenerife accident* (Technical Report). Washington, DC: Air Line Pilots' Association.

Rosenthal, U., & Pijnenburg, B. (1991). Simulation-oriented scenarios. An alternative approach to crisis decision making and emergency management. In U. Rosenthal & B. Pijnenburg (Eds.), *Crisis management and decision making*. Netherlands: Kluwer Academic.

Rosenthal, U., Comfort, L. & Boin, A. (in press). (Eds). *Managing crises: Threats, dilemmas, opportunities*. Springfield, IL: CC Thomas.

Roth, E., Mumaw, R. & Lewis, P. (1994). *An empirical investigation of operator performance in cognitively demanding simulated emergencies* (NUREGCR-6208). Washington, DC: Division of Systems Research, Office of Nuclear Regulatory Research, US Nuclear Regulatory Commission.

Rouse, W., Cannon-Bowers, J., & Salas, E. (1992). The role of mental models in team performance in complex systems. *IEEE Transactions on Systems, Man, and Cybernetics, 22*, 1295–1308.

Rubinstein, E. (1979). Special Issue on Three Mile Island. *IEE Spectrum, 16*.

Salas, E., Bowers, C., & Edens, E. (2001). (Eds.) *Improving teamwork in organizations: Applications of resource management training.* Mahwah, NJ: Lawrence Erlbaum Associates.

Salas, E. & Klein, G. (2001). (Eds.) *Linking expertise and naturalistic decision making.* Mahwah, NJ: Lawrence Erlbaum Associates.

Schmitt, J. (1994). *Mastering tactics. Tactical decision game workbook.* Quantico, VA: Marine Corps Association.

Schmitt, J. & Klein, G. (1996). Fighting in the fog: Dealing with battlefield uncertainty. *Marine Corps Gazette, 80,* (August), 62–69.

Serfaty, D. & Michel, R. (1990). Toward a theory of tactical decision making expertise. Paper presented at Symposium on Command and Control, Monteray, CA.

Skriver, J., & Flin, R. (1996). Decision making in offshore emergencies: Are Standard Operating Procedures the solution? In *Proceedings of SPE International Conference on Health, Safety and Environment in Oil and Gas Industry, New Orleans, Louisiana.* (pp. 443–447). Richardson, Tx: Society of Petroleum Engineers.

Skriver, J., & Flin, R. (1997). Emergency decision making on offshore installations. In D. Harris (Ed.), *Engineering psychology and cognitive ergonomics* (Vol. 2 – Job design and product design, pp. 47–54). Aldershot: Avebury.

Stern, E. (1999). Crisis decision making: A cognitive institutional approach. Unpublished PhD. Department of Political Science, University of Stockholm.

Stokes, A. (1997) Aeronautical decision making. In C. Zsambok & G. Klein (Eds.) *Naturalistic decision making.* Mahwah, NJ: Lawrence Erlbaum Associates.

Stokes, A., & Kite, K. (Eds.). (1994). *Flight stress Stress, fatigue, and performance in aviation.* Aldershot: Avebury.

Svenson, O. (1997). Differentiation and consolidation theory. In R. Flin, E. Salas, M. Strub, & L. Martin (Eds.), *Decision making under stress. Emerging themes and applications.* Aldershot: Ashgate.

Taylor, P. (1990). *The Hillsborough stadium disaster. Final Report.* London: Home Office.

Tissington, P., Flin, R., & McGeorge, P. (1998). Decision making by UK firefighters. Paper presented at the Fourth Conference on Naturalistic Decision Making, Warrenton, Virginia.

US Marine Corps. (1999). *Command and Control* (MCDP 6 (PCN 142 000001 00)) Quantico,VA: US Government.

van Avermaete, J. & Kruijsen, E. (1998). *NOTECHS: The evaluation of non-technical skills of multi-pilot aircrew in relation to the JAR-FCL requirements* (Project Report CR 98443). Amsterdam: NLR.

VECTOR Command UK. (1999). Vector Command Training Simulation (Version 02/06/00): www.vectorcommand.co.uk.

Wickens, C. (1996). Designing for stress. In J. Driskell & E. Salas (Eds.), *Stress and human performance*. Mahwah, NJ: Lawrence Erlbaum Associates.

Yates, F. (2001)."Outsider": Impressions of naturalistic decision making. In E. Salas & G. Klein (Eds.), *Linking expertise and naturalistic decision making*. Mahwah, NJ: Lawrence Erlbaum Associates.

Zakay, D., & Wooler, S. (1984). Time pressure, training, and decision effectiveness. *Ergonomics, 27*, 273–284.

Zsambok, C. (1997). Naturalistic decision making: Where are we now? In C. Zsambok & G. Klein (Eds.), *Naturalistic decision making*. Mahwah, NJ: Lawrence Erlbaum Associates.

Zsambok, C. & Klein, G. (Eds). (1997). *Naturalistic decision making*. Mahwah, NJ: Lawrence Erlbaum Associates.

12 Command and Control Teams: Principles for Training and Assessment

Eduardo Salas, Janis A. Cannon-Bowers* and Jeanne Weaver
University of Central Florida
*Naval Air Warfare Center

Eduardo Salas is a Professor of Psychology at the University of Central Florida and Principal Scientist for human factors research at their Institute for Simulation and Training. He is the Director of UCF's Ph.D. Applied Experimental and Human Factors Psychology Program and Editor of the Human Factors Journal. He is an Honorary Professor of Psychology at the University of Aberdeen. Previously, he was a senior research psychologist and Head of the Training Technology Development Branch of the Naval Air Warfare Center Training Systems Division for 15 years when he served as principal investigator for R&D programs focusing on teamwork, team training and decision-making under stress.

Janis A. Cannon-Bowers, Ph.D, is senior research psychologist in the Science and Technology Division of the Naval Air Warfare Center Training Systems Division (NAWCTSD), Orlando, Florida. As the team leader for advanced surface training research Dr. Cannon-Bowers has been involved in research projects directed toward improving training for complex environments. These have included investigation of training needs and design for multi-operator training systems, training effectiveness and transfer of training issues, tactical decision making under stress, the impact of multimedia training formats on learning and performance, and training for knowledge-rich environments.

Jeanne Weaver is an Assistant Professor at the University of Central

Florida Department of Psychology.   Her Ph.D is in Human Factors Psychology and her research interests are stress, human performance (individual and team), and medical error.

The study of command and control teams is a relatively new area of psychological research. However, perhaps due to the critical nature of the problems investigated, researchers dedicated to the study of this topic have made concerted and persistent efforts to advance knowledge in this area. In fact, Cannon-Bowers and Salas (1998a) state that while the issues in this field are far from fully resolved, much has been done in the last decade to derive principles for training *and* to alter the degree of attention given to issues surrounding human performance in command and control settings. However, Cannon-Bowers and Salas (1998b) note that although the research in this area has 'exploded,' there is still much to be done to further our understanding of command and control operations.

The impetus for the work described here was largely the catastrophe of the USS *Vincennes*. The *Vincennes* was a Navy AEGIS system, guided missile cruiser. The purpose of the craft was to protect aircraft battle groups against air attack. Tragically, on July 3, 1988, a number of factors converged that led to the destruction of a commercial Iran Air flight that killed 290 people. Specifically, the crew of the *Vincennes* arrived at the decision to shoot down the airbus based on the ambiguous data available in a situation already characterized by uncertainty and volatility. Although in retrospect it is easy to see that the decision was clearly the wrong one, on July 3rd the correct course of action was far from clear. It quickly became evident that the factors that contributed to the disaster were in dire need of scientific scrutiny.

Consequently, the purpose of the current chapter is to summarize the research conducted on training and assessment in command and control teams. First, however, we will provide a context for the chapter by defining the characteristics of the problem. Next, we will review some of the most pertinent research regarding training. We will then review existing research regarding command and control measurement and assessment. Each of these sections will end with specific guidance for trainers/assessors of command and control teams. Because of the nature of the command and control environment, the review sections will necessarily incorporate relevant studies regarding training and assessment under stress conditions. It should be noted that the research described herein has all been conducted in the last ten years. In addition, a substantial portion of the research described was conducted under the umbrella of the Tactical Decision Making Under Stress (TADMUS) project (see Cannon-Bowers & Salas,

1998a). The chapter will conclude with a prescriptive research agenda with which to further knowledge in this area.

The Nature of Command and Control

This section will briefly describe the characteristics common to the command and control (C&C) environment and define what is meant by the term 'team' in this context. The command and control environment is one in which optimal decisions must be made in situations characterized by severe time pressure, a high degree of uncertainty and ambiguity, and conditions that change and evolve in a rapid fashion (Cannon-Bowers & Salas, 1998a). Furthermore, the task at hand requires multiple people to gather, process, integrate, and perform based on the data accumulated. Given these circumstances, it is readily apparent that command and control is a situation characterized by stress. Consequently, the literature regarding C&C teams almost inevitably involves the mention of 'stress.' For the purposes of such team task performance situations, stress has been defined as 'a process by which certain environmental demands…evoke an appraisal process in which perceived demand exceeds resources and results in undesirable physiological, psychological, behavioral, or social outcomes (Salas, Driskell, & Hughes, 1996, p. 6). Those who would seek to optimize performance in these settings must acknowledge the inherent stressful nature of command and control.

Finally, it is necessary to define the term 'team.' One of the most oft used definitions of this term is that provided by Salas, Dickinson, Converse, and Tannenbaum, (1992). Their definition states that a team is at least two or more people who are working toward a shared and valued goal. The definition emphasizes that the individuals' efforts must be interdependent. That is, in order for successful task performance to occur, the individuals' efforts must be coordinated. The interdependent nature of the command and control task likewise adds to the challenge for those who attempt to train and assess the performance of C&C teams. Consequently, the derivation of guidelines that address each of these issues with regard to training is of paramount importance.

It has also been emphasized by researchers within this area that it is important to acknowledge that there are two tracks of behaviors required for successful team task performance. The two tracks of behaviors in which team members must acquire and demonstrate competency are teamwork and taskwork. According to McIntyre and Salas (1995) *taskwork* refers to those behaviors that are related to the operations-related activities that team members must perform. These comprise the technical aspects of the team task to be performed. In contrast, *teamwork* behaviors are those that act to 'strengthen the quality of functional interactions, relationships, cooperation, communication, and coordination of team members' (McIntyre & Salas, 1995, p. 15). Thus, taskwork refers to those behaviors that encompass the team's interactions. Table 12.1 provides a description of those teamwork skills determined to be most critical and definitions of each.

Table 12.1 Teamwork dimensions and definitions (from McIntyre & Salas, 1995)

	Dimensions	Definition
Teamwork Behaviour	Performance Monitoring	Team members observe the behavior of team members and accept that their behavior is being observed or monitored as well.
	Feedback	A climate in which feedback is freely offered and accepted among team members.
	Closed-Loop Communication	A three step sequence whereby a message is sent by a team member, another team member provides feedback regarding the received message, and the originating team member then verifies that the *intended* message was received.
	Backing-Up Behaviors	Behavior among team members that indicates both the competence and willingness to help, and be helped by, other team members.
Team Performance Norms	Team Self-Awareness	The attitude or value that team members see themselves first as members of the team and as individuals second.
	Fostering of Team Interdependence	A team member attitude that reflects the extent that individual team members' success depends on the success of other team members as well.

As shown in Table 12.1, McIntyre and Salas (1995) argue that there are four critical teamwork behaviors and two team performance norms. The dimensions listed here are interdependent. For example, feedback would obviously occur with performance monitoring in an effective team. Furthermore, it is noted that these behaviors occur in a dynamic context. Figure 12.1 depicts the Team Effectiveness Model that includes teamwork skills as one of the contributing factors to team performance.

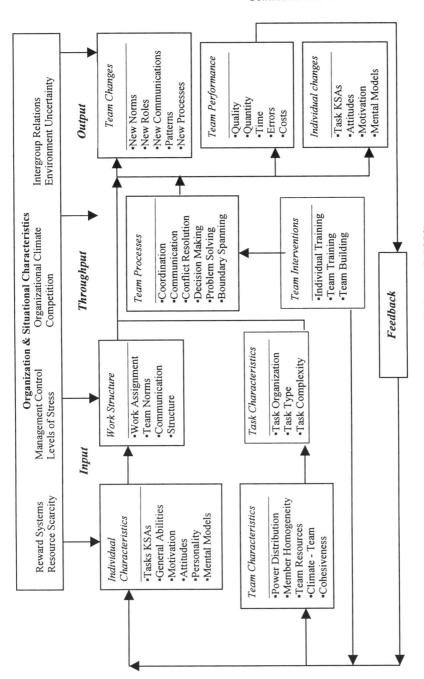

Figure 12.1 Team Effectiveness Framework (adapted from Salas et al, 1992).

This comprehensive model includes a number of other factors that contribute to the overall effectiveness of a team's performance as well. This model exemplifies the complexity of the team task performance situation and it is certainly no less complex for teams in a command and control environment.

Training in Command and Control

As previously noted, training in the command and control environment offers numerous challenges to researchers. Characteristics that must be considered in the development of training for this operational environment include the nature of the task and environment, the fact that teams are the trainees of interest, the interdependence of the tasks to be performed, and the inherent stress of the situation. Therefore, the TADMUS program sought to consider all of these factors in the systematic test and development of training methods for use in this arena. Specifically, Cannon-Bowers and Salas (1998a) sought to extend knowledge in this area by grounding the development of training approaches in the C&C environment in existing theory. These theories included shared mental models, automaticity, naturalistic decision making, goal orientation, meta-cognition, and stress inoculation. In fact, these authors took the a priori position that by generating hypotheses from existing theory, their potential for success would be greatly increased.

Having laid the theoretical foundation, a number of hypotheses to be tested were generated. Among these were that training should: 1) be event-based, 2) focus on teamwork skills, 3) provide team leaders with particular skills, 4) provide team members with interpositional knowledge, 5) teach team members to self-correct with regard to performance, and 6) prepare teams to cope with the stressful nature of the task environment. The global goals corresponding to these methods of training were to increase overall performance readiness, train stress coping skills, and to focus on team and individual skills that would be most vulnerable to stressor effects (Cannon-Bowers & Salas, 1998a). The fulfillment of these goals was believed to be most likely attainable by focusing on the identification and testing of the *content* of training. Consequently, the section that follows will summarize some of the research conducted under the TADMUS program in order to ascertain what methods or content would prove most effective in eliciting effective team performance in these conditions. The review is organized into three general areas. These areas correspond to training of team *members,* team *leaders*, and team *interactions.* Obviously there is some overlap between these in terms of training focus, but it should be noted that the intent of all of these interventions is to improve team process/performance.

Team Member Training

Volpe, Cannon-Bowers, Salas, and Spector (1996) conducted an experiment in order to test the extent that cross-training might mitigate the effects of workload. Cross-training can be defined as a strategy that instructs team members in the duties of their teammates. Such training includes information regarding the performance of the roles and tasks of the other team members as well as provision of information regarding the overall team framework. The researchers manipulated two levels each of workload and cross-training in order to assess the effects on task performance, communication, and teamwork. The participants were 40, two-person teams composed of undergraduate students. The subjects were asked to perform a simulated flying task whereby they were to 'shoot down' enemy aircraft. Performance dependent variables were such behaviors as number of enemy craft shot down, time to destroy craft, and the number of times enemy craft were allowed to lock on with their radar. Process measures consisted of assessment of teamwork behaviors such as communication and interpersonal cooperation. The findings of the study indicated that teams who received cross-training had better teamwork than teams without cross-training and cross-trained teams also used more efficient communication than non-cross-trained teams. Furthermore, cross-trained teams also were more effective in terms of performance. However, cross-training did not interact with workload in the manner expected. Workload did significantly impact teamwork and communication in the expected direction. Although this study does indicate the general utility of cross-training for team performance, the findings are less clear with regard to its effectiveness in mitigating the effects of workload.

A replication and extension of the Volpe et al. study was conducted by Cannon-Bowers, Salas, Blickensderfer, and Bowers (1998) utilizing a task requiring more interdependence among team members. Specifically, the task was a simulated radar scope that required team members to make decisions regarding the intent of incoming targets presented on the radar screen. Measures of performance included number of targets engaged and time to process information regarding incoming targets. Another measure of interest in this study was the degree of interpositional knowledge (IPK) possessed by team members. Although this study utilized a different task and three-member teams of Navy recruits, the results were also supportive of the utility of cross-training. The teams who received the cross-training intervention had performance superior to the non-cross-trained teams. Specifically, the cross-trained teams engaged more targets correctly and were faster with regard to handling the target information available. Cross-trained teams also volunteered more information to their team mates in agreement with the findings of the Volpe study. However, unlike the Volpe study, Cannon-Bowers et al. (1998) also found an interaction between cross-training and workload with regard to team score and overall team process quality. Specifically, teams that received cross-training were able to maintain their performance and process under high workload relative to teams who did not receive cross-training. However, under low workload the teams did not differ with regard to performance or process measures. Furthermore, teams that were cross-trained exhibited higher levels of IPK than non-cross trained teams.

Smith-Jentsch, Salas, and Baker (1996) investigated the effects of assertiveness training. They note the proposition of other researchers that assertiveness is critical in effective team decision making. Three ideas tested by Smith-Jentsch and her colleagues (1996) were that team-related assertiveness differs from general assertiveness with regard to their specific classes of behaviors, the interpersonal context influences the assertiveness response individuals exhibit, and team performance-related assertiveness consists of both an attitudinal and a skill component. The results of this three-part investigation indicated that directiveness and independence subscales on a subjective measure of assertiveness were better predictors of team performance-related assertiveness than social assertiveness and defense of interest subscales. It was also found that team performance-related assertiveness was superior for those who received assertiveness training than for those who did not. However, training that used lecture and demonstration that *also* emphasized practice and feedback was superior to assertiveness training that did not include the opportunity for active practice. Based upon the results of their studies, these researchers concluded that changes in assertiveness were a function of skill development and not attitude change. Furthermore, because their results indicated that assertiveness differs depending upon the context, selection of persons for work team situations should account for these contextual differences in assertiveness levels. Finally, the subjective dimensions of assertiveness most closely related to team performance-related assertiveness were directiveness and independence. Thus, these would be the dimensions to target most closely in training interventions. Although assertiveness was not determined to be a critical factor in the *Vincennes* incident, it is nevertheless an important component of team process and thus has been widely studied in aviation (e.g. Orasanu, 1999; see also Lodge, this volume).

A recent study by Inzana and her colleagues (Inzana, Driskell, Salas, & Johnston, 1996) investigated the utility of the use of preparatory information as a training method for reducing stress reactions and facilitating performance on a realistic decision making task. Prior research regarding the utility of such information provision has been conducted almost exclusively in health care settings. There are three types of preparatory information that have been described (Driskell & Johnston, 1998). *Sensory information* communicates how the person is likely to feel when exposed to stressors. This includes physiological and emotional reactions. *Procedural information* describes the stressor environment that will be encountered and includes information regarding the types of stressors and the setting itself. *Instrumental information* communicates to the individual what should be done in order to cope with the negative aspects of the stress response.

The participants in the study were 92 Navy enlisted personnel from the Naval technical school. Performance accuracy and speed were both assessed as dependent measures on a task that simulates Naval decision making. That is, subjects were to identify, label, and dispense with each target that appeared on the radar scope in accord with information regarding its type, status, and intent. Subjects' subjective stress and self-efficacy were also assessed. The results of the study were as follows: individuals provided with preparatory information prior to performance in a task related stress environment reported less anxiety, had more confidence in their ability to perform the task (i.e., self-efficacy), and had superior performance in terms of errors than those who were not provided with such information. However, these researchers suggest that future work investigating the utility of preparatory information for minimizing stress in task performance situations focus on 'determining the relative effects of these specific types of information' (p. 434).

Team Leader Training

Another approach has targeted training team leaders as a method for improving overall team performance (Tannenbaum, Smith-Jentsch, & Behson, 1998). Tannenbaum and his colleagues (1998) argue that team leaders can improve the performance of their teams by acting as a team *facilitator*. That is, rather than acting in a traditional supervisory manner, team facilitators work with their teams in order to develop ways to improve performance, further the team's development, and assist the team in learning as a team. Tannenbaum et al. (1998) sought to investigate the extent that a) team leaders could be taught to conduct more effective pre-briefs and post-action reviews, b) team members would act differently based upon changes in their leader's behaviors, and c) improvements in team and leader briefing behaviors would influence team performance. Participants in this study were 70 naval officers who were recruited to form 14, five-person teams that were required to perform a simulated, networked Combat Information Centre task. The primary performance measures were time and accuracy with regard to target identification. Process measures included the assessment of briefing and teamwork behaviors. Significant findings regarding the hypotheses were as follows. First, trained team leaders did in fact exhibit more effective briefing behaviors than the control group. This finding was particularly visible in the post-action reviews vs. the pre-briefs. Leaders who had been trained were more likely to encourage team member involvement by asking team members to critique themselves and to guide the team to consider teamwork behaviors. These leaders also provided more positive feedback, probed more, asked for more feedback, and were more likely to engage in self-critiquing behaviors. Thus, team leaders can indeed be trained to display effective briefing behaviors.

With regard to the second research question, although team members did not act differently during pre-briefing, teams with trained leaders did display different behaviors during post-action reviews. In particular, team members with trained leaders were more likely to engage in self-critiques and to offer suggestions to their team mates. They were also significantly more likely to take part in discussions about teamwork behaviors. In summary, training team leaders does appear to create a climate in which team members felt better able to express themselves.

Finally, the results regarding team leader training and performance were positive as well. Not only did teams with trained leaders exhibit more effective teamwork behaviors but such teams also achieved better performance outcomes. That is, the teams' assessments of critical scenario events were more accurate and timely than those made by teams with untrained leaders. Overall, team leader training was successful with regard to each of the three key research questions.

Smith-Jentsch, Zeisig, Acton, and McPherson (1998) contributed to the acquisition of knowledge regarding training in their consideration of 'team dimensional training' (TDT), a method for teaching teams to self-correct their behaviors in order to improve the development of their teamwork-related knowledge and skills. In particular, the purpose of TDT is to facilitate the development of shared mental models. These mental models have been defined as knowledge structures or cognitive representations that people use to organize newly accumulated information. Furthermore, not only do they use these models to describe, explain, and predict events (Rouse & Morris, 1986) but they also use them to guide their interactions with other people (Gentner & Stevens, 1983). Thus, Smith-Jentsch and her colleagues argue that team leaders and trainers can contribute to the development of these shared mental models. This is the purpose of TDT.

This work also sought to extend the earlier work described below in the measurement section conducted by Smith-Jentsch, Johnston, and Payne (1998). The TDT utilized both classroom training and hands-on practice. The classroom portion was approximately four hours long, while the hands-on practice portion ranged in time from four to eight hours depending upon the size of the class. The results of the training were positive as indicated by a card sort task that assessed the extent to which the newly trained instructors possessed mental models in line with that of the TDT model. There was a significant change in the instructors' mental models following training. In addition, data from instructors who were exposed only to the classroom training was compared with that of instructors who experienced the practice session as well. Instructors who took part in the hands-on training accumulated significantly more knowledge than those who received only classroom instruction. The authors concluded that TDT could help maintain the focus of team discussions, improve team climate, increase team member participation, and improve the utility of team member input.

Training Team Interactions

The final training method to be discussed is a team training strategy called team adaptation and coordination training (TACT) (Serfaty, Entin, & Johnston, 1998). The primary purpose of this training method is to improve team coordination performance. In order to assess the utility of TACT, an experiment was conducted comparing TACT with a control condition and a condition with TACT plus situation assessment updates by the tactical action officer. A stressor manipulation was also included as an additional manipulated variable. The participants were twelve, five-person teams composed of active duty military officers. These teams were tested utilizing a simulated C&C environment and measured in terms of both their processes and performance. The outcome measures consisted of twelve behaviorally anchored items related to performance (e.g., whether radar detections were reported and communicated in a timely manner). The measure of process was the Teamwork Observational Form that assesses the dimensions of team orientation, communication, monitoring, feedback and coordination. The results indicated that the TACT intervention had a significant effect on the composite of the twelve behaviorally anchored performance items. Specifically, performance of teams receiving the TACT training and TACT plus tactical action officer input performed better following exposure to the intervention *and* outperformed teams in the control condition. In addition, teams performed better during stressor exposure following the intervention. With regard to teamwork, the pattern of results mimicked that of the performance findings with trained teams, exhibiting better teamwork skills following training and better skills compared to the control group. The teams' communications in the trained conditions were also changed in a pattern consistent with the training. The authors concluded that the team training strategy was successful and its simplicity makes the training of practical utility for onboard training systems.

There are a number of other useful publications regarding training (cf. Salas, Cannon-Bowers, & Blickensderfer, 1995; Salas, Cannon-Bowers, & Johnston, 1997). However, it is beyond the scope of this chapter to review all of these manuscripts. Instead we have tried to focus almost exclusively on those training test and development efforts with the most direct applicability to the command and control environment. The section that follows draws upon the research cited previously, however many of these principles likely would apply to other environments as well.

Training Principles

This section offers a number of principles for the guidance of training development and implementation based upon the research cited above (see also Cannon-Bowers et al., 1995). These principles vary from strategies for improving team performance by increasing the skills of the team leader to improving the coordination of the teams' members. It is hoped that these principles will provide direction for trainers of command and control teams.

Principles that Target Team Members

Cross-training (cf. Volpe, Cannon-Bowers, Salas, & Spector, 1996; Cannon-Bowers, Salas, Blickensderfer & Bowers, 1998)

- Trainers should assess cross-training needs and develop the appropriate intervention for meeting these needs by utilizing data obtained via team task analysis.
- The primary goal of cross-training is to provide team members with an understanding of their teammate's responsibilities. Thus, such training must necessarily demonstrate what behaviors are required for other members of the team *and* when these behaviors would be appropriate.

- The provision of cross-training for teams is valuable because it facilitates team members' capacity to anticipate the needs of their teammates and thus improves their ability to provide resources and information as they are required.
- Cross-training interventions should provide practice opportunities for team members in conducting the tasks of the others and provide feedback based upon their practice sessions.
- When cross-training is delivered it should focus not only on the task requirements but address the coordination or process of the team members as well, particularly for highly interdependent team tasks.
- The interdependence of the team task must be considered in the development of cross-training interventions. The higher the inter-dependence among members, the stronger the need for cross-training.

Preparatory Information (cf. Inzana, Driskell, Salas, & Johnston, 1996)

- Preparatory information should be used to enhance performance under stress.
- Preparatory information should be used to increase self-efficacy in task performance situations characterized by stress.
- Provision of preparatory information should be used to reduce subjective perceptions of stress in task performance situations characterized by stress.

Assertiveness Training (cf. Smith-Jentsch, Salas, & Baker, 1996)

- Trainers should utilize practice and feedback in order to develop task-related assertiveness.
- Assertiveness training for team task situations should emphasize specific training of independence and directiveness.
- Because assertiveness differs depending upon the context, selection and training of persons for work team situations should account for these contextual differences.

Principles that Target Team Leaders and Team Processes

Team Leader and Team Process Training (cf. Tannenbaum, Smith-Jentsch, & Behson, 1998; Smith-Jentsch, Zeisig, Acton, & McPherson, 1998)

- Trained team leaders will conduct more effective briefings than team leaders who receive no specialized training.
- Team leader training contributes to the situational awareness of the team leader.
- In order to facilitate team learning, it is important to conduct pre and post briefings.
- Team coordination skills can be trained as well as task-related skills.
- Team coordination training can be achieved relatively simply using instruction, demonstration, practice, and feedback.
- Shared mental models must be developed *but* use of these models must also be trained to be adaptable to the dynamics of situational demands.
- Event-based training should be used whenever practicable.
- Team feedback should emphasize processes rather than outcomes.
- Critical incidents should be built into the development of scenarios in order to elicit team supportive behaviors.

Assessment and Measurement in Command and Control

It has been argued that measurement in C&C must take a four-pronged approach. This method considers both process and outcomes at the team and individual levels (Cannon-Bowers & Salas, 1997). First, measures of performance (MOPs) are measures of process that assess the strategies and procedures that accompany task performance. Furthermore, Cannon-Bowers and Salas argue that these measures must be obtained at both the team and individual level. Second, measures of effectiveness (MOEs) are outcome measures that should also be attained for teams and team members. In order to develop and assess training it is necessary for process measures to be gathered along with outcome measures in order to provide effective feedback to teams. Smith-Jentsch et al. (1998) summarize this perspective nicely with the statement that 'whereas process measures are critical for providing feedback to trainees, outcome measures are needed to identify which processes are, in fact, more effective.' (p. 63)

Smith-Jentsch et al. (1998) describes the development and evaluation of a human performance measurement system with the purpose of assessing training needs. These authors argue that this measurement system should describe, diagnose, and evaluate the processes that are associated with effective outcomes. As stated above, such a system must consider both processes and outcomes or MOPs and MOEs. The specific purpose of Smith-Jentsch and her colleagues was to determine whether the tools for performance measurement developed within the TADMUS program met these criteria. In particular, two principles were emphasized. First, remediation should emphasize processes that are linked to outcomes and second, individual and team level deficiencies should be distinguishable in order to support the instructional process.

The approach taken by Smith-Jentsch and her colleagues was as follows. The foundation of the approach rested upon the notion that event-based techniques would be emphasized in order to acquire measures of both team and individual processes that could then be linked to critical outcomes. Thus, this event-based approach first requires the a priori identification of the events of interest. The events are specifically selected in order to allow the manifestation of behaviors consistent with the learning objectives of the scenario. Both process and outcome measures are obtained at each event and these measures then serve as data with which to identify those behaviors that are associated with effective outcomes such as timeliness of track identifications, assessments of intent, and decisions to utilize weaponry.

The analysis of this data yielded some 'lessons learned' based on this approach. Some of the team process measures yielded poor reliability and high intercorrelations. A couple of reasons were cited for this finding. First, reliability was particularly low for those dimensions that required raters to infer a state of mind vs. those that allowed the observation of overt behavior. Second, there was some redundancy across the seven rated dimensions. Consequently, later efforts narrowed the original seven dimensions to four. These dimensions were information exchange, supporting behavior, communication, and team initiative. Although a complete description of the analyses utilized to assess the utility of these four dimensions is beyond the scope of this chapter, (see Smith-Jentsch, Johnston, & Payne, 1998 for a complete description) it is possible to summarize their findings based upon these analyses.

The composite ratings of the four identified dimensions were indicative of superior teamwork strategy. Teams with more experience received higher ratings than less experienced teams. The information exchange, team initiative, and communication dimensions were each uniquely related to the quality of team decisions. This occurred across distinct scenario events. Although this research did extend knowledge regarding team performance measurement, the authors suggest that future research should attempt to determine the relationship between individual and team performance indices.

Assessment principles (cf. Smith-Jentsch, Johnston, & Payne, 1998)

- Ratings of a team's performance should be obtained via observation. That is, team assessment should identify and then measure those behaviors necessary for effective team performance.
- In determining the 'most important' dimensions of teamwork necessary for effective performance, it should be considered that 'less is more.' That is, the fewest number of dimensions accounting for the highest degree of variance in team performance should be striven for.
- Although identification of the most critical dimensions of teamwork is important, *actual assessment* should be conducted by measuring *specific behaviors* within dimensions.
- Utilizing an event-based approach to assessment is important. For example, it provides direction regarding the relationship between process and outcome, and allows direct knowledge regarding the occurrence or lack thereof of critical team behaviors.
- The assessment of team training should be conducted in a dynamic manner.
- Team measurement should be conducted at the team and individual levels. Such assessment provides data necessary to diagnose deficits in team processes and/or performance.
- Team assessment data should be descriptive of the teams' behaviors as well as providing for the evaluation of performance.
- Team assessment should be developed not only to evaluate the quality of performance but also to allow the diagnosis of deficiencies in order to direct remedial training.
- Whenever practicable, team assessment should occur at the behavioral, cognitive, and attitudinal levels.

Summary and Research Agenda

The research that has been discussed here represents a systematic program of research from which stems valuable guidance for training and measurement of command and control teams. In general, this research represents efforts to develop training targeting team members, team interactions, and team leader behavior. It is fair to say that the literature here is very useful in terms of provision of information that can be further tested and applied in yet other environments. For example, event-based approaches are a critical component of team training in the C&C environment and would likely be effective in other environments as well.

Further research could test the extent to which the principles detailed here might apply to other operational team environments (e.g. surgical and medical teams). In summary, event-based training has been shown to be a critical component of team training for command and control. Likewise, the research that has been conducted provides support for the utility of behavioral measures and yet points out the need for advances in measurement with regard to measuring team knowledge structures.

Baker and Salas (1997) have noted that future research should focus on the development of measures that assess interpositional knowledge. They state that this is a critical issue because as teams improve, there can be a loss in observable team behaviors. Consequently, assessment instruments must be able to capture processes that occur beyond the typical purview of measurement devices that exist today.

Cannon-Bowers and Salas (1998a) also provide guidance with regard to future directions for research. These authors offer three areas in need of further research. These areas relate to the investigation and establishment of techniques for analyzing team tasks, team cognition, and improving methods that will facilitate the ability of teams to maintain continuous learning (e.g., team self-correction and feedback). Cannon-Bowers and Salas (1998a) argue that the capability must exist for team tasks to be analyzed. In particular, they note that it is important to be able to identify the knowledge, skills, and abilities required by team tasks. Another area recommended for further research is the study of team cognition. It is important to further our understanding of the manner in which teams arrive at points of shared knowledge. Finally, teams must be taught to maintain attitudes and behaviors consistent with continuous learning. There is a need to further explore ways to encourage such team development.

Research must also continue in order to identify ways to assess responses on multiple levels to stressors in applied environments. At the same time, it is also critical to understand how these responses might inter-relate and the extent that they are related to a team's processes and performance. Future studies regarding the utility of preparatory information could attempt to determine the types of preparatory information that are most beneficial for moderating negative performance outcomes in the face of various stressors.

Although the research that has been conducted with regard to the utility of team leader training has been very promising, future efforts could explore still other behaviors that leaders might learn in order to facilitate their teams; processes and performance. In addition, future work could also study further the assertiveness concept and ways to insure timely and appropriate assertiveness behavior while attempting to link these behaviors to positive team performance outcomes.

In summary, a great deal has been learned in a relatively short period of time. However, the critical tasks faced by command and control and other types of operational teams contribute to a continued sense of urgency with regard to furthering our understanding of the best ways to train and assess team processes and performance. By using the research of the last decade as a springboard, it is possible that *our* knowledge structure regarding team training, assessment and performance can be expanded in an increasingly expedient manner. However, as previously noted, given the increasingly cognitive nature of the tasks teams are expected to perform, it is likely that the magnitude of the challenge will remain.

References

Baker, D. P., & Salas E. (1996). Analyzing team performance: In the eye of the beholder? *Military Psychology, 8,* 235–245.
Cannon-Bowers, J. A., & Salas, E. (1997). A framework for developing team performance measures in training. In M.T. Brannick, E. Salas,, et al (Eds.). *Team performance assessment and measurement: Theory, methods, and applications.* (pp. 45–62). Mahwah, NJ: Lawrence Erlbaum Associates.
Cannon-Bowers, J. A., & Salas, E. (1998a). Individual and team decision making under stress: theoretical underpinnings. In J.A. Cannon-Bowers, & E. Salas. (Eds.), *Making decisions under stress: implications for individuals and teams* (pp. 17–38). Washington, DC: American Psychological Association.
Cannon-Bowers, J., & Salas, E. (1998b). Team performance training in complex environments: Recent findings from applied research. *Current Directions in Psychological Science,7,* 83–87.
Cannon-Bowers, J. A., Salas, E., Blickensderfer, E. & Bowers, C. A. (1998). The impact of cross-training and workload on team functioning: A replication and extension of initial findings. *Human Factors, 40,* 92–101.
Cannon-Bowers, J. A., Tannenbaum, S. I., Salas, E., & Volpe, C. E. (1995). Defining competencies and establishing team training requirements. In R.A. Guzzo, E. Salas & Associates (Eds.) *Team effectiveness and decision making in organizations* (pp. 333–380). San Francisco: Jossey-Bass Publishing.
Driskell, J. E., & Johnston, J. H. (1998). Stress exposure training. In J.A. Cannon-Bowers, & E. Salas. (Eds.) *Making decisions under stress: Implications for individual and team training.* (pp. 191–218). Washington, DC: American Psychological Association.
Gentner, D., & Stevens, A. I. (1983). *Mental Models.* Hillsdale, NJ: Lawrence Erlbaum.
Inzana, C.N., Driskell, J.E., Salas, E., & Johnston, J.E. (1996). Effects of preparatory information on enhancing performance under stress. *Journal of Applied Psychology, 81,* 429–435.
McIntyre, R.M. & Salas, E. (1995). Measuring and managing for team performance: Lessons from complex environments. In R.A. Guzzo, E. Salas, & Associates (Eds.) *Team Effectiveness and Decision Making in Organizations* (pp. 9–45). San Francisco: Jossey Bass.
Orasanu, J. (1999) Has CRM succeeded too well? Assertiveness on the flightdeck. In R. Jensen (Ed.) *Proceedings of the Tenth International Symposium on Aviation Psychology.* Columbus: Ohio State University.

Rouse, W. B., & Morris, N.M. (1986). On looking into the black box: Prospects and limits in the search for mental models. *Psychological Bulletin, 100,* 349–363.

Salas, E., Cannon-Bowers, J. A., & Blickensderfer, E. (1995). Enhancing reciprocity between training theory and training: practice, principles, and specialties. In J. Ford, S. Kozlowski, K. Kraiger, E. Salas & M.S. Teachout. (Eds.). *Improving training effectiveness in work organizations.* (pp. 291–321). Mahwah, NJ: Lawrence Erlbaum Associates.

Salas, E., Cannon-Bowers, J. A., & Johnston, J. (1997). How can you turn a team of experts into an expert team: emerging training strategies. In C.E. Zsambok & G. Klein. (Eds.) *Naturalistic Decision Making.* (pp. 359–370). Mahwah, NJ: Lawrence Erlbaum Associates.

Salas, E., Dickinson,T. L., Converse, S. A., & Tannenbaum, S. (1992). Toward an understanding of team performance and training. In R.W. Swezey, & E. Salas, (Eds.). *Teams: Their training and performance.* (pp. 3–31). Norwood, New Jersey: Ablex.

Salas, E., Driskell, J. E., & Hughes, S. (1996) The study of stress and human performance. In J. E. Driskell, & E. Salas, (Eds.) *Stress and human performance.* Mahwah, NJ: Lawrence Erlbaum Associates.

Serfaty, D., Entin, E. E., & Johnston, J. (1998). Team coordination training. In J.A. Cannon-Bowers & E. Salas. (Eds.), *Making decisions under stress: implications for individuals and teams.* (pp. 221–246). Washington, D.C.: American Psychological Association.

Smith-Jentsch, K. A., Johnston, J. H., & Payne, S. C. (1998). Measuring team-related expertise in complex environments. In J.A. Cannon-Bowers, & E. Salas (Eds.), *Making decisions under stress: implications for individuals and teams.* (pp. 61–87): Washington, DC: American Psychological Association.

Smith-Jentsch, K. A., Salas, E., & Baker, D. (1996). Training team performance-related assertiveness. *Personnel Psychology, 49,* 909–936.

Smith-Jentsch, K., Zeisig, K. L., Acton, B, & McPherson, J. A. (1998). Team dimensional training: a strategy for guided team self-correction. In J. A. Cannon-Bowers, & E. Salas, (Eds.), *Making decisions under stress: implications for individuals and teams.* (pp. 271–298). Washington, DC: American Psychological Association.

Tannenbaum, S., Smith-Jentsch, K., & Behson, S. J. (1998). Training team leaders to facilitate team learning and performance. In J.A. Cannon-Bowers, & E. Salas, (Eds.) *Making decisions under stress: Implications for individuals and teams* (pp. 247–270). Washington, DC: American Psychological Association.

Volpe, C.E., Cannon-Bowers, J.A., Salas, E., & Spector, P. E. (1996). The impact of cross-training on team functioning: An empirical investigation. *Human Factors, 38,* 87–100.

13 Training Decision-Making by Team Based Simulation

Jonathan Crego and Claire Harris*
Metropolitan Police Service, New Scotland Yard
*University of Plymouth

Dr Jonathan Crego is Director of Learning Technologies at the Metropolitan Police Service in London. He is responsible for Critical Incident Management training for senior officers. He designed both the MINERVA and HYDRA immersive simulation systems. He received his Doctorate at Salford University for the design of team-based learning environments.

Claire Harris is an undergraduate student in the Department of Psychology at the University of Plymouth. She worked with Dr Crego during a placement at the operations facility at Bramshill Police Staff College, Hampshire.

www.calt.demon.co.uk
Jcrego@calt.demon.co.uk

This chapter describes the use of simulation to generate a highly credible training environment in which senior decision-makers can be immersed. A sophisticated system called Hydra used by the Metropolitan Police, will be described. Its evolving design (and that of an earlier system called Minerva, see Crego & Spinks, 1997) is driven by both an organisational need for enhanced personal accountability in the decision-making process and to accredit decision-makers as competent.

The police service has, over the last ten years, been endeavouring to enhance the professionalism of critical decision-makers. The Taylor Report (1990) produced after the Public Inquiry into the Hillsborough Football

Disaster, was a catalyst for research and development. The report was highly critical of the competence demonstrated by senior officers during the management of a critical incident, in 1989 where 95 football fans lost their lives due to a crowd-crushing incident in the ground. In an analysis of similar Public Inquiries over the last eighty years (Crego, 1996), a clear pattern emerges. Senior command teams, faced with the complexities of an evolving incident, encounter an often overwhelming flow of information, are surrounded by uncertainty, as well as conflicts of meaning, and have to operate within very short time frames for proactive intervention.

Recommendations contained within these Inquiries unfortunately share a similar theme. Not only is the competency of those who are placed in a position of command questioned, but the recommendations also suggest that training for these officers should be designed to properly prepare them for the chaotic world of critical incident management.

The Chaotic World of Critical Incident Management

This chapter attempts to address the difficulties faced by senior decision-makers during the management of critical incidents. What tools are there to help them? It would seem that there is no general decision-making model that completely describes the decision making process at critical incidents to act as a guide. Therefore we open with a discussion around the term 'critical incident' and suggest that its meaning be inexorably linked to the environment decision-makers find themselves operating within.

Surely to pursue a Holy Grail, seeking a universal theory that can be applied to the myriad of operational situations labelled as 'critical', is destined to end in frustration and disappointment? What is necessary, is an approach that selectively integrates those elements of differing theories and applies them to the critical incident environment under examination. In other words picking the best from a multitude of theoretical models.

We then go on to describe an immersive simulation system called Hydra, designed to provide senior strategic decision-makers with a feeling of 'being there' or presence as the key decision-maker at a critical incident. This simulated environment provides a useful observational tool by which to monitor real-time decision-making, to provide directed feedback to those who may be called upon to take on the role of incident commander.

The final part draws conclusions from observations of senior police decision-makers in a Hydra exercise and discusses the overlapping, yet polarising, theories of Classical Decision Making (CDM) and Naturalistic Decision Making (NDM).

Critical Incidents

Instantaneous Versus Protracted Incidents: Fast and Slow Burners

Critical incidents have many different interpretations. For example, the senior management team at *Perrier*, the mineral water company, on finding that their product had become contaminated with benzene, decided eventually to recall 160 million bottles, after mismanaging the opening phase of the crisis (see Seymour & Moore, 2000). The 'golden hours' approach to crisis management focuses on the importance of decision making in the critical initial response phase. This should increase the likelihood that the immediate disaster is managed decisively, (or at least the external perception of decisive action) and can contain the impact of the initial triggers. This approach is successfully applied to Public Disorder and Public Safety Incidents within the police service and the fire service, where immediate saving of life and minimisation of property damage are prioritised. First we need to define the term 'critical incident'.

- Just what is a critical incident?
- When does an incident become critical?
- Where are the boundaries?

For example, from a lay perspective, a major crushing occurring at a street carnival looks like a critical incident. Delivering food to hostage takers after the delivery has been agreed and negotiated by the hostage negotiator team, in discussion with firearm advice, looks like a critical incident. Dealing with community leaders over many months during an ongoing inquiry into a racial murder looks like a critical incident. Managing a fire on an oilrig looks like a critical incident. Giving authority to deliver a deliberate assault by armed police officers at a siege looks like a critical incident. Dealing with the melt down of a nuclear reactor looks like a critical incident. Managing the slaughter of 600,000 infected cows with mad cows' disease (BSE: Bovine Spongiform Encephalopathy) looks like a critical incident. Clearly military pilots in the fighter jets would have a different perception of the term 'critical' when applied to an enemy radar lock rather than to their perception of a hostage negotiation. It would seem that the word 'critical' must have a number of meanings and those meanings are related to the reference frame of the expert group (see also Quarantelli, 1998).

A blunt tool for differentiation of incidents is the element of time. Much is written about the impact of time and its ability to influence decision-making. Taking the radar lock example above, a short time for response may be thirty seconds, the difference between life and death. The number of options available to the flight crew may be limited but the time available will undoubtedly influence their thinking. Classical decision making research (see Connolly, Arkes & Hammond, 2000) has provided many models and prescriptions for selecting the optimal course of action to respond to a given problem. Naturalistic decision making research (NDM: Zsambok & Klein, 1997) focuses on experts making decisions in real world environments. These are typically characterised by time pressure, high risk, dynamic problems and inadequate information and there are a number of NDM decision models that may be applicable in critical incidents. For example Klein's (1998) Recognition Primed Decision-Making (RPD) describes within high stakes and a short time frame for action, the retrieval of patterns of typicality within the critical incident that the decision makers recognise from their previous experiences and their selection of a course of action becomes a rapid process. But does time matter? Do the 'slow-burners' (incidents that have a much longer duration) present the same problems for the decision-maker? The continuing environmental impact after the fire at Chernobyl generated a critical incident that has lasted for many years, with ever-greater complexity as time moves on. A murder investigation lasting several months is not a critical incident in one sense but in another, it is a collection of critical incidents, a number of slow-burners that become as Rittel and Webber (1973) would suggest 'wicked problems' – problems that change even as a function of their observation.

So where does this leave us, – more confused? Ready for even more complex overarching decision making models, that attempt to bridge the gap between different types of decision making for critical incidents that are both fast and slow-burners.

Developing Decision Makers

Whether looking at fast or slow-burning critical incidents, competing decision making theories cloud the horizon. Getting novices to think like experts requires the engendering of experience. Much has been debated during psychology conferences and in papers about experience-led decision making, recognising patterns within the complexity of an unfolding set of problems (e.g. Klein, 1998). Whole new schools of thinking have emerged within this arena but there is a need to reconcile the various models and approaches from competing decision theories, especially NDM vs. CDM (see Lipshitz et al, in press; Yates, (2001), for recent expositions of this debate).

Successful decision-makers seem to follow a pattern where their decisions are based on the recognition of the problem at an early stage, this is the essence of Klein's RPD model (see Klein, 1998). However, worked through mental analyses that take in all the possibilities of the problem are more typical of the analytical or Classical Decision-Making approach (Janis & Mann, 1972) and this also requires an experience base. So both RPD and CDM are in agreement that experience is a key element.

It is to the engendering of experience that this paper now turns.

The Hydra Experience

To provide a learning environment in which senior decision-makers can become immersed in a set of problems has been the driving factor of the design of the Hydra system. This has been built on research carried out using the Minerva simulator now installed at:

- London Fire Brigade
- New South Wales Police Service Australia, (used to train all the event commanders for the Sydney Olympics 2000)
- Police Staff College Bramshill, Hampshire
- South Wales Police Service.

The Hydra system has been developed to meet the needs of decision-makers operating in slow-burning environments, such as a murder investigation or a hostage negotiation. It requires decision making at a strategic as well as a tactical level (see Arbuthnot, this volume). Students are split into three syndicates (of 5–7 members) and each syndicate operates within a separate room known as a 'pod'. Each syndicate has to manage the evolving critical incident, simultaneously working through the exercise from the same starting event. This common starting point ensures that the plenary sessions, which occur throughout the exercise, are tremendously rich as each syndicate may well have approached and managed the problems differently. The syndicate rooms are equipped with computer screens, keyboards and printers that are controlled by a central control room, operated by three subject matter experts. The Hydra syndicate control software enables these experts to respond to the syndicates' calls for information. The dynamic nature of the exercise requires that the world the Senior Investigating Officers (SIO) of each syndicate inhabit is continually changing. Each participant in turn takes on the role of SIO. The incident unfolds in real-time with information becoming available through a knowledge base of possible outcomes. When the students are immersed in their syndicate rooms, this knowledge base helps subject matter experts generate responses to their specific management of the incident as well as general information that becomes available as time progresses, delivered by audio reports, printer messages and video footage. Consequently, the world may unfold differently for different syndicates as a direct result of their interventions. The system is designed to respond directly to the syndicates and therefore at each plenary session, different syndicates will have reached different conclusions and will have recorded different decisions on their Decision Log. This is a tool for recording on the computer not only the decision but also the rationale behind its formation. The record of these decisions and rationales can be projected onto a large screen for the whole group to discuss at the plenary session. Discussion about the rationale for the decision forms a major part in the learning outcome.

Figure 13.1 Example of the Hydra Decision Log

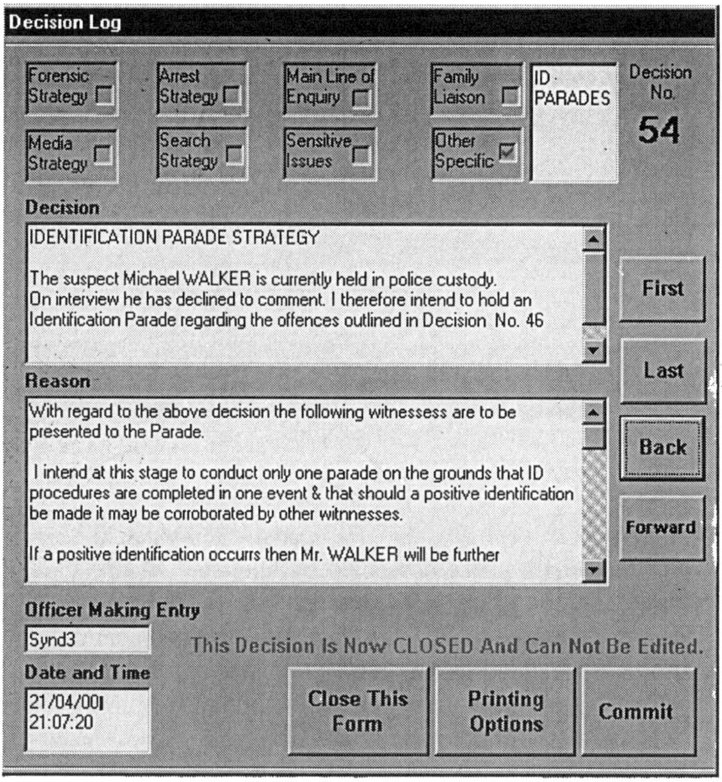

The Hydra system needs to be able to expose information to the students in an incremental manner, either as a result of decisions being made by the syndicate or as a function of the unfolding enquiry. A key element of the interface is therefore part of the computer system called the Communicator.

Figure 13.2   Example of the Hydra Communicator

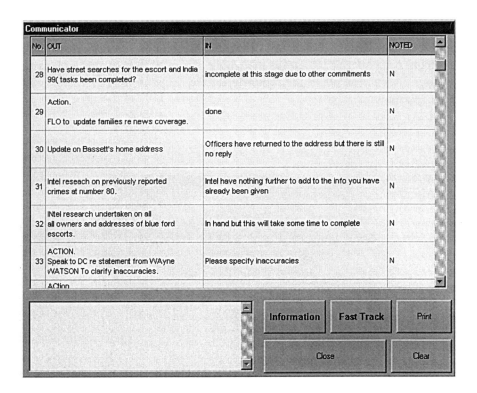

The Senior Investigating Officers are able to intensively search for new information (akin to the search phase described by Janis and Mann, 1972) by using the Communicator. This interface acts as a repository for information as it becomes known to the enquiry team. Decisions and their resulting information are reflected in this repository through the increased availability of such information as statements, photographs, exhibits and intelligence. The system also provides video and other multimedia cues in the same manner as the Minerva system (see Crego & Spinks, 1997). The screen shot below shows an example of such an interaction.

Figure 13.3 Video and Audio Repositories in Hydra

Observing Decision Making within the Simulated Environment

A number of Senior Investigating Officers were observed participating in a Hydra exercise. The exercise was centred on the investigation of a complicated murder of an elderly Asian couple and the linked double abduction and rape of two 'special-needs' young girls. The exercise forms a major part of the British National Crime Faculty's Senior Investigating Officer Development Program, called the Management of Serious Crimes Course. Officers taking part in the program spend three weeks at the Detective Training School, followed by a modular course of legal, forensic, procedural and intelligence elements. They then spend three weeks attached to a Major Incident Pool, which is the organisational resource for the investigation into murder and other serious crimes. After completing this attachment they return to the Training School and are immersed in an exercise running within the Hydra complex. This final part of the training lasts for five days and prospective Senior Investigating Officers have to complete this phase before they can be accredited with the title of Senior Investigating Officer and by definition, investigate murders.

The authors felt that the Hydra immersive element would provide a useful vehicle through which they could observe the Senior Investigating Officers engaged in real-time decision making without impacting on the special and fragile dynamics of the group. This observational method was easy to accomplish as each of the syndicate rooms was equipped with CCTV and boundary microphones. The Hydra control room, by design, monitors syndicate activities and this monitoring was made available for the research.

This observational approach was used to confirm our view that decision-makers are continually demonstrating elements of both analytical (CDM) and recognitional (RPD) processes when making decisions (see Crichton and Flin, this volume for a more detailed description of these decision processes). Many of the problems they faced required a team-based approach to problem solving which is traditionally associated with more analytical methods (CDM). However, there were many instances where their behaviours could be better described by RPD, particularly when making split second judgmental decisions.

Videotapes of the sessions were analysed to identify decision conversations between the SIOs. The exercise ran for sixty hours over five days. During this time the incident continuously moved between slow and fast-burn. For example, careful analysis of the witness statements, together with examination of forensic evidence, was considered against a backcloth of highly volatile family liaison, press involvement, ministerial and chief officer distractions. To support our argument that no overarching model truly describes the decision making that occurred during these periods we have selected the following examples.

An example of the participants displaying both recognitional and analytical decision methods was when they had to make a decision as to whether or not to allow an officer to enter the house of the elderly Asian couple. They had conflicting information to deal with (i.e. one neighbour said that they had not seen them for a while and that they were worried, whereas another said that the couple had gone away for a few days). The SIOs had a strong time pressure to cope with in this situation (i.e. preservation of life, as they were unsure if the old couple were still alive).

These characteristics are typical of NDM environments (time pressure, high stakes, uncertainty, multiple players, and organisational goals, Klein, 1997) where one might have expected to find the police officers using a fast RPD style of decision making. However they were observed to be weighing up the pros and cons of the decision in a more classical or analytical fashion – *'we should go in...', '...but if we do and it's wrong...we just have to patch it up and apologise...'* Eventually a decision was made to enter the house and this decision was recorded using the Decision Log. The individuals used an analytical method presumably because they had time to do so and were attempting to optimise in order to find the best course of action. Moreover they knew that later they would have to justify their choice of decision to a complex situation where the preservation of life is the most important factor.

When dealing with the suspect, the students were desperately looking for more information from their 'Intelligence Cell' on the suspect before they were in a position to obtain a search warrant to enter his property. There was a strong time pressure to catch the offender, as not apprehending the criminal with the likelihood that he would re-offend, produced extremely high stakes. The group discussed the problem, calling for advice from team members, they then, through conversation, generated a situational awareness of a mental map or model of how to go about sealing off the crime scene. This team approach required the use of mental templates/models (through schemata, which drive the search for information) of past experiences and 'simulation heuristics' (Kahneman & Tversky, 1982) to move along the path from the 'problem state' to a 'solution state' (Newell & Simon, 1972).

When in discussion about whether or not to arrest the principal suspect for the murders there is, again, evidence to suggest that the participants seamlessly transcended between two decision making methods, clearly demonstrating behaviours attributed to recognitional as well as analytical processes. One SIO argued that the whole team needed to discuss the pros and cons for arresting him straight away; they decided (in an analytical mode) that they needed to establish whether or not he was on the DNA database *'...it takes four days... I've had to do this... then nick him again two weeks later...'* This evidence suggests that the decision makers were using a recognition-primed approach; they were, in the short term, using satisfycing goals of their past experiences to form a concept of this particular event so as to investigate it properly.

Our observations of the SIO exercise, coupled with many hundreds of hours of experience simulating critical incidents, clearly identifies the overlap that exists between recognitional and analytical styles of decision making, applied to fast and slow-burn incidents. What is becoming clear is that the mutually exclusive labelling of decision making as Classical or Naturalistic is unsafe and if exercises are developed and designed around this artificial divide, the needs of the trainees will not be met.

Future Research

It is possible that the whole approach to the development of critical incident management may be over-simplified. Much research has been carried out within the emergency services and other similar environments (petro-chemical, nuclear, military etc.), learning from disaster and adopting recommendations emanating from Inquiries, investigations, criminal and civil actions (see Cannon-Bowers & Salas, 1998; Flin 1996). The findings are fed-back into the design and development of policy and training. But the emerging question put simply is this:

If these officers, managers, pilots or whoever had their time again, would they perform any better? We think they would!

The theoretical analysis of critical incidents is in danger of over-simplification.   Reductionalist approaches distil down to the smallest element, trigger or cue patterns emerging from chaos and attempt to attach meaning or significance or even worse some unifying theory. One is left with an unpleasant feeling of realisation that critical incidents are, (as best put by Rittel 1973) 'Wicked Problems'.

Our approach is to continue to develop training exercises that reflect the complexity of the real world and provide many opportunities for critical decision-makers to learn from their experience within the high fidelity world of Hydra.

References

Cannon-Bowers, J. & Salas, E. (1998) (Eds.) *Making Decisions Under Stress.* Washington: APA Books.
Connolly, T., Arkes, H. & Hammond, K. (2000) (Eds.) *Judgement and Decision Making (2nd ed.).* Cambridge: Cambridge University Press.
Crego, J.   (1996). *Critical Incident Management: Engendering Experience Through Simulation.* PhD Thesis Salford University.
Crego, J. & Spinks, T. (1997) Critical incident management simulation. In R. Flin, E. Salas, M. Strub & L. Martin (Eds.) *Decision Making under Stress.* Aldershot: Ashgate.
Flin, R. (1996) *Sitting in the Hot Seat: Leaders and Teams for Critical Incident Management.* Chichester: Wiley.
Janis, I. & Mann, L. (1977).   *Decision Making – A Psychological Analysis of Conflict, Choice and Commitment.* New York: Free Press.
Kahneman, D. & Tversky, A. (1982). The simulation heuristic. In D. Kahneman, P. Slovic, A. Tversky (Eds.) *Judgment Under Uncertainty.* Cambridge: Cambridge University Press.
Klein, G. (1997)   Recognition Primed Decision Making.   Where are we now?   In C. Zsambok & G. Klein (Eds) *Naturalistic Decision Making.* Mahwah, NJ: Lawrence Erlbaum Associates.
Klein, G. (1998). *Sources of Power: How People Make Decisions.* Cambridge: MIT Press.

Lipshitz, R., Klein, G., Orasanu, J. & Salas, E. (in press) Taking stock of NDM. *Journal of Behavioural Decision Making.*

Newell, A. & Simon, H. (1972). *Human Problem Solving..* Englewood Cliffs, NJ: Prentice Hall.

Quarantelli, H. (1998) (Ed.) *What is a Disaster?* London: Routledge.

Rittel, H. & Webber, M. (1973). Dilemmas in a general theory of planning. *Policy Sciences, 4,* 155–169. (Republished as 'Planning problems are wicked problems' In N. Cross (Ed). (1984). *Developments in Design Methodology* . Chichester: Wiley).

Seymour, M. & Moore, S. (2000) *Effective Crisis Management.* London: Casell.

Taylor, P. (1990) *The Hillsborough Stadium Disaster. Final Report.* Home Office, London: HMSO.

Yates, F. (2001) 'Outsider': Impressions of Naturalistic Decision Making. In E. Salas & G. Klein (Eds.) *Linking Expertise and Naturalistic Decision Making.* Mahwah, N.J.: Lawrence Erlbaum Associates.

Zsambok, C. & Klein, G. (1997) (Eds.) *Naturalistic Decision Making.* Mahwah, NJ: Lawrence Erlbaum Associates.

14 Lessons from the Hot Seat

Rhona Flin and Kevin Arbuthnot*
University of Aberdeen
*West Yorkshire Fire Service

Introduction

This final chapter summarises the central themes emerging from the previous cases and research summaries. Our contributors have addressed, either directly or indirectly, the essence of incident command and how the component skills can be developed and maintained. We review our authors' opinions on four key issues in an effort to establish a consensual position on the current role and anticipated challenges for the incident commanders of the 21st Century. Our selected themes are:

- The Essence of Command
  - Leadership and Team Management
  - Stress Management
  - Situation (Risk) Assessment
  - Decision Making
- Developing Commanders
  - Selection
  - Training
  - Competence Assessment
- The Command Organisation
  - Incident Command Systems, Tiers of Command
  - Communication
- Political, Media and Legal Considerations.

The final section concludes with an examination of future directions and a list of outstanding questions.

Throughout this chapter we have drawn on McCann and Pigeau's (2000) excellent volume on military command, *The Human in Command*, which was published last year. They set out with an aim, similar to our own, of examining the nature of command by combining personal accounts of military commanders' experience with reviews of relevant research from military psychologists.

The Essence of Command

All the commanders who have contributed to this volume have discussed the essential nature of command, either explicitly or implicitly. The term 'Command and Control' was explored in Chapter 2, but it was recognised that some of the definitions may have become worn or stilted over time (see also Flin, 1996, p. 20). Not surprisingly, our authors' definitions of the term 'commander' and of the nature of command are determined by their particular occupational background – the commander of a civilian aircraft with an in-flight emergency is dealing with a set of demands and conditions which are distinctive from the fireground commander or the military commander. Yet from a close reading of their experiential accounts, it is possible to discern a consensus on what they consider to be the essence of command, namely that commanders are trusted with the authority and responsibility to organise, direct, co-ordinate and control resources to achieve a given aim. In the case of incident commanders, the objective is usually to manage and resolve an emergency situation. Likewise, Crabbe (2000, p. 10) discussing the nature of command, cites a NATO definition as *'authority vested in an individual .. for the direction, co-ordination and control of military forces'*, but offers his own wider interpretation to encompass responsibility, accountability, the setting of moral standards and key decision making/judgmental skills.

Whichever phrasing is adopted as the definition of command, the similarities in our authors' accounts lie in the contextual factors and their identification of the core skills required to take charge of a critical incident. Sarna summarised the context of the incident commander's task with the definition, 'Critical incidents…are life safety events'. To achieve some distinction between the concept of the commander, as opposed to that of the manager or leader, we proceed with the assumption that the commander is: 'The leader in a dynamic, safety critical environment'. This would resonate well with Fuller's belief cited by Arbuthnot, that in essence, a commander is only commanding when influencing operations, and exercising the authority that has been vested in him or her.

So what are the fundamental skills of an effective commander? Sarna offers a detailed list which includes: conducting situation assessments, making high-risk decisions, co-ordinating the activities of numerous specialists, negotiating with other agencies, prioritising, handling multiple problems, and managing personal stress levels. Leadership and team management are mentioned by every commander in this volume. From our reading of the preceding chapters we have formulated a taxonomy of core incident command skills as follows:

(i)   Leadership and Team Management
(ii)   Stress Management
(iii)   Situation (Risk) Assessment
(iv)   Decision making.

In comparison, McCann and Pigeau (2000) summarising the views of military commanders, group command capabilities into four categories: Intellectual (cognitive skills of information gathering, situation assessment, problem solving and decision making); Emotional (resolve, resiliency, adaptability, retention of perspective and a sense of humour); Interpersonal skills (leadership); Team Building capabilities (form teams and foster group effort). Our categorisation overlaps that of McCann and Pigeau (2000) with the intellectual (cognitive) skills separated into two distinct skill categories, and the interpersonal (social) skills of leadership and team management discussed under the same heading. We would agree that attributes such as resilience, resolve and a sense of humour are highly desirable command characteristics and our authors would endorse this view. However with the exception of stress management, these would appear to constitute facets of personality, which may have to be favoured in the selection process rather than command skills, which can be formally trained. Our contributors' views on the four sets of essential skills are summarised below.

Leadership and Team Management

All the contributors have emphasised leadership as central to the command role, stressing the importance of team building and personal credibility throughout. This is independent of whether the organisation is military, emergency service or industrial. The requirements of good planning, team co-ordination, resource management, firm and visible direction/guidance and, of course, achievement of objectives are virtually indistinguishable beneath the surface features of their personal accounts.

Those who have not had experience of exercising command are sometimes misled by the hierarchical appearance of organisational charts and line relationships. The autocratic language of published orders, procedures and policies in the military or emergency service environments compounds this to a great extent. Unfortunately it leads to distorted beliefs gaining currency about what 'command and control' actually is. In reality the relationships between 'leader and led' in such organisations are much more complex, two-way and consensual. Questions of effective leadership, especially in modern organisations functioning as part of contemporary society, must be considered as functions of the effectiveness of the team as a whole, and not simply of the formal leader or 'boss' in isolation.

The success of any venture is arguably as much dependent upon the 'followership' of those being led as it is upon the display of classical leadership traits in the leader. This particularly applies to an empowered workforce or team, which operates within a framework of delegated authority. Despite common misconceptions, this is very much the situation in the organisations under consideration in this volume. Soldiers, police officers or firefighters must be able to operate with considerable autonomy in the field, and are trusted to exercise good judgement and be effective without supervision where necessary, e.g., when the supervisor is incapacitated or when communications break down. To facilitate this, organisational policies and mission objectives must be unambiguous and command structures fluid. It is in this complex and uncertain environment that the effective commander must be able to flourish.

The elements of leadership that involve team building, mutual support, effective delegation and trust are all of critical importance to personnel working in dynamic, high risk situations. Not surprisingly, they are emphasised by the armed services' commanders who stress the importance of presenting the human face of the commander to those being led (see also Cherrie, 2000). Acting as a positive role model to the team and leading by example, as well as building mutual respect, appear to feature consistently in the traits that experienced senior commanders consider essential for success. This is another manifestation of the importance of symbolism in command. Symbolism, whether referred to explicitly or not, has been a feature in all of the accounts of the exercise of command. Powerful images have been generated of courageously leading from the front (Larken), participation in physical training with the troops (Keeling), or the conscious nurturing of the image of the 'old man' in the command chair (Coyle).

A commander's presence extends beyond the physical, however. There are issues associated with team-building and team identity that are dependent to a large extent on the commander's personality and charisma. This is the commander's presence in a non-physical sense, which inspires 'followership' and secures the commander as the head of, as well as part of, the team being led. The commander's whole demeanour is arguably an aspect of 'presence'. Whatever the organisation or environment, a leader who displays self-control and continues to function effectively in moments of high drama gains respect. It is understandable, therefore, that in some situations, particularly when the commander's crews have been exposed to danger, that the commander has felt drawn towards the scene of action. There is clearly some degree of symbolism here. Apart from the functional benefits or otherwise of doing this, the commander is consolidating his or her credibility and credentials with a clear demonstration of solidarity with the team in the crisis. This will without doubt assist the commander on future occasions and probably constitutes part of the commander's developmental process if done proportionately.

There are also considerations of the team members', and others', expectations and needs to which the commander must be both alert and able to meet. Keeling gave a useful account of how he saw his responsibilities to others during his Belfast deployment, which might be summarised as follows:

(i)   To his seniors, where he felt a duty to establish good relations and gain respect.

(ii)   To his counterparts and peers in other units and services whose close co-operation and empathy would be crucial to success of the mission.

(iii)   To be an effective leader to his team, (which is the part of the role that has been explored more than others in treatises on leadership and command).

This describes a role that has responsibilities in all directions and effectively portrays how, in the chain of command, each commander's position is critical in information flow and support of the ongoing task. It is an expression of the art of team management that ranges across several possible definitions of the word 'team'.

Team management is an obvious and necessary extension of the concept of leadership. As with any individual, a team has an identity, requires briefing, praising, disciplining and it must bond to be effective (see Salas et al). The leader must be part of the team and not separate from it. As Keeling has effectively shown, there are various concepts of 'team', ranging from the small units working closely together as flight deck crews or army platoons, to larger and more fluid groupings across whole organisations or events. Each type has its needs and peculiarities, which have been covered variously in this volume. Team members sometimes have to manage the boss as well as their peers and subordinates. It is not a concept inextricably linked with hierarchies, however, for practical reasons, in larger teams, hierarchies tend to develop and operate as a necessary feature of the structure.

Stress Management

Brunacini reflected on his experience of a pipe-smoking role model who never exhibited excitement or panic. The composure of the commander is certainly an essential ingredient of credibility, which the majority of contributors to this volume have referred to as an essential quality or characteristic. It is difficult to fake being calm if one is not, although the issue is complicated by the fact that some serious disorders of decision making can be masked by what might appear to be composure. Calmness and composure would appear to have a symbolic value as much as a practical one. Panic is infectious. The experienced commander is acutely aware of that fact and will, accordingly, take care to present the right image as well as do the right things. Sarna touches on the same idea by expressing that 'calm is contagious'. Presenting a calm demeanour will be all the more readily achievable if the commander has had the benefit of exposure to stressful conditions in controlled situations such as exercises and has gained experience of dealing with it. Keeling is firm in the view that'... training exercises should be hard and should cause you to make mistakes ... (or) ... you will not learn the demands of real incidents'. Lodge describes 'learning from panic' and Sarna agrees, adding the view that testing simulations will prepare incident commanders, and particularly novices, '..to manage performance degrading stress effects in themselves and others effectively'. This is, therefore, a clear steer for trainers.

Before leaving the subject of stress, it is worth noting two further points. First, that incident commanders need to be acutely aware of the longer-term risks of stress to those under their command (Dallaire, 2000). Post-traumatic stress syndrome (PTSD) is a debilitating and damaging injury. It can also result in expensive litigation if the commander or the organisation is found to have been negligent. There is now ample evidence that the risks of PTSD can be minimised by appropriate protective and responsive measures during and after exposure to high level traumatic situations (Eid, 2000; Yule, 1999).

Second, commanders themselves are not immune to the effects of being in a demanding or exciting situation. Moore commented that he was not conscious of any 'increased adrenaline flow' until after the event when he experienced a sense of euphoria after the event had been successfully concluded. Reactive effects are not always positive and commanders need to realise, especially in longer running situations, that as well as remaining alert for signs of stress in others, they need to monitor their own level of stress (Keeling, this volume; Woodward, 1992). This is important because, as both Coyle and Larken point out, 'command is a lonely business' and senior commanders may be isolated from both their peers and their own managers. Everts (2000), a Colonel in the Royal Netherlands Army dealing with a difficult humanitarian mission in Sebrenica, describes asking two of his battalion officers to watch over him and to warn him if he was going 'off the rails'. The military commanders also emphasised the risks of fatigue and the importance of 'pacing yourself' in long-run missions. This can be pertinent to commanding certain policing (e.g. hostage taking) and industrial emergencies where an emergency situation may take days before control is regained. While there will be inherent personality differences in resilience, there are certainly stress management techniques that can be trained (see Driskell & Salas 1996; Flin, 1996)

Situation (Risk) Assessment

This volume contains several personal references to the criticality of situation assessment as an aspect of the skill of exercising command. Whether this has been raised in the context of 'keeping one's eye on the ball' or of the difficulties a commander experiences in building a picture of a developing incident, the issue is fundamentally one of information assimilation and analysis in tight time frames. The overlapping concepts of situation assessment and situation awareness (the former being the process used to build the latter) are now of mainstream interest to psychologists studying performance in high reliability settings, such as aviation and anaesthesia (see Endsley & Garland, 2000; Orasanu, 1995; Fletcher et al, in press). Investigations of mid-ranking incident commanders' decision making in the fire service and in industry are also confirming the primacy of situation assessment in the decision making process (Crichton, in prep; Tissington, 2001).

Risk assessment is sometimes used as another way of expressing the concept of situation assessment, but with a focus on some future action or process that is being contemplated. Risk assessments are conducted in business dealings or even domestic life at least as frequently as they are in hazardous operations. Nevertheless, when considering critical incident decision making, risk assessment is one of the most common judgements that commanders make. This should be set as part of a robust process and should include an agreed default path.

In the West Yorkshire Fire Service the dynamic risk assessment process formalises a 'default to defensive' mode for operations where the hazards and risks can not immediately be quantified (for example, at hazardous materials incidents or complex technical rescues from machinery). What this means in its simplest form is that if the on-scene commander is not satisfied that he/she has a suitable and sufficient awareness of the situation to permit a crew to operate in the hazard zone (or in 'offensive mode'), then he/she must refrain from immediately deploying them in the affected area. A 'defensive' approach must be taken until such time as an analytical (i.e., conducted in a classical decision making style) risk assessment has been conducted, and all hazards identified, control measures put in place and suitable safe systems of work employed. This may seem obvious and certainly experienced commanders have traditionally employed such an approach, but the formalisation and teaching of it ensures that the wisdom and rationale is communicated to the less experienced at a much earlier stage in their development. However experienced the individual or team may be, every commander has a personal threshold between the familiar and the unfamiliar. An experienced commander will need to resort to default modes less frequently than a junior colleague because of more experience and training.

Psychological research in the area of situation assessment/awareness is still at a relatively early stage but the accumulated evidence to date indicates that the skill components can be identified and trained (Endsley & Garland, 2000; Fallesen, 2000; Salas et al, 2001). The aviation industry leads the field in this area (Lodge refers to the new training course being used by *British Airways* to heighten pilots' situation awareness). The Royal Aeronautical Society organised a conference on the subject in May 2001 which was attended by representatives from the emergency services, medicine and the energy sector industries, as well as the airlines, indicating the growing awareness of this subject.[1]

Decision Making

Naturalistic Decision Making (NDM) is a relatively new domain of psychological research which examines the ability to make decisions in real world conditions, typically in potentially demanding environments (e.g.

aviation, armed forces, medicine, emergency services) (see Flin et al, 1997; Salas & Klein, 2001, as well as Sarna, Larken, Crego, and Crichton, in this volume). Clearly this is one field of scientific research which the practitioners feel has potential for their operations. Within the NDM arena, different methods of decision making are used by expert commanders: these include systematic evaluation of optional courses of action (often called analytical decision making); formal rules/procedures; and intuition-like processes (e.g. recognition-primed decision making (RPD). A number of our contributors (Larken, Sarna, Crego) acknowledge the importance of this intuitive, 'gut feel' decision making. This is echoed by the military commanders in McCann and Pigeau's book. For example, Cherrie (2000) discussing senior US army commanders says 'they could all look at a complex problem and know, intuitively, exactly what needs to be done' (p. 19). Crabbe (2000) agreed, stating that military commanders 'must be able to think clearly, based on experience, and make those 'gut feel' decisions calmly and accurately' (p. 12). Also Labbe (2000) believes that commanders should have, 'both sufficient professional knowledge of the conduct of operations and sufficient field experience to exercise effective and timely military judgement based on an informed understanding of the situation – in short a recognitive quality referred to as *intuition*.'(p. 111). He also mentions German generals attributing differences in speed of decision making to *fingerspitzengefuhl* – 'fingertip feeling'.

In contrast to the formal, classical forms of decision making, we have seen that RPD is an intuitive process, which provides a commander with a decision with minimal conscious deliberation, but is it always the right decision? It is well recognised that prompt decisions are more likely to be effective in dynamic circumstances than slow ones, but is there any compensation for, or acknowledgement of, the fact that some will be flawed, and of those, some will be very flawed? There are clearly limits to the reliability of RPD decisions, but how can that threshold of integrity be raised? Because of the stakes in critical situations it is necessary to ensure that decision makers are trained to deal with situations beyond, as well as within, their capabilities. They must recognise when they are faced with a situation that is too rapidly developing or too complex to deal with in the ways and tempos they had previously been familiar with.

Despite the emphasis that has been placed on the significance of the fast, intuitive styles of decision making in this volume, this must not be taken to imply that there is no place for analytical styles. An over-reliance on naturalistic styles of decision making could lead to problems in the training and development of commanders. The psychologists have accepted that the RPD process is dependent upon valid experience and training. The form of decision making that is taught in basic training is necessarily analytical in style. Basic operations and manoeuvres are taught to military personnel, firefighters and emergency medics, and drilled repeatedly to ensure that those who employ them in safety critical circumstances know them inside out and will not stumble due to stresses caused by demands of time or danger, and with good reason. This is the foundation upon which a commander's later quality decisions will be built. It is the integrity and underpinning of any system that accepts that commanders must, at times, make rapid, intuitive style decisions and that they will be reliable. It goes without saying that it is critical that departures from formal procedure or standard behaviours are done from a position of knowledge and experience of those systems rather than ignorance or incompetence.

Every commander who has written in this volume has included reference to planning, including explicit maxims such as 'to fail to plan is to plan to fail'. The maxim is no less true for being familiar. Preparation has featured in the Royal Marines, Royal Navy, UK and US police and the prison commanders' stories. It is highly relevant to the fire service, with a duty in the UK to collate tactical and strategic information on all high risks for use when the emergency occurs. This is where the basis of all effective decision making lies. During the planning phase there should be time to gather all available information and to consider the merits of alternative courses of action. Commanders who devote appropriate time to planning their operations will receive an excellent return on investment when faced with their next incident.

Developing Commanders

Selection

The commanders actually chose to say very little on the subject of their selection into senior command roles, despite being encouraged to do so. Their brief biographies show that they all rose rapidly through the ranks of their profession, having entered it at an early stage. Possibly they are too modest to discuss their own attributes and the skills that afforded them their promotions to senior command positions. In fact there is a very limited literature on desired characteristics of commanders and the relative merits of selection techniques (Flin, 1996). For some domains such as aerospace, policing, or military (see Flin, 2001) detailed empirical accounts of selection tests and methods are available, but they tend to concentrate on *ab initio* selection and rarely focus on command attributes. There are some valuable anecdotal accounts to draw on. Based on his experience in the US army, Cherrie (2000) a retired Brigadier outlines the key characteristics of senior commanders. These are vision, intelligence, competence, above average mental and physical stamina, a strong sense of mission awareness, and immense self-confidence. With candour, he adds a complementary list of negative attributes, which he believes are typical of senior commanders. The 'dark side' characteristics are – limited caring skills, obsession with perfection, an unwillingness to handle situations involving risk, and micromanagement. This is an interesting profile and underlines that there may be traits that organisations would wish to 'select out' as well as desirable characteristics to be 'selected in'.

At present a wide range of different personality measurement instruments are being used for selection of commanders (Flin, 2001), and there is no one recommended test for command selection. Current research on personality in the workplace (Roberts & Hogan, 2001) has been focussing on what are called the Big Five personality characteristics and while there are suggestions that extroversion and emotional stability are potential predictors of command competence, there are insufficient data to draw robust conclusions. Other recent findings are beginning to re-assert the importance of cognitive skills in command-type occupations (Buck, 1999; Fallesen, 2000). Given the requirement for situation assessment and decision making, then better selection techniques for identifying individuals who have or can acquire key cognitive skills would seem to be desirable.

Many high reliability domains favour the use of simulators for identifying suitable command candidates. However, some recent studies (Howard & Choi, 2000) have shown that very low fidelity selection techniques such as situational interviews (where a candidate's description of how they would solve a given problem, is scored against expert answers) are showing promising predictive properties for selection.

Training

The commanders all underlined the importance of training, but several mentioned that there used to be very limited if any specific, formal training on command skills. Keeling said that in his military career he had been given, 'no formal advice on command', and likewise Moore comments on the absence of decision training for police commanders. Lodge felt that learning command skills from role models on the flight deck had been 'a hit or miss affair' and Coyle noted that some senior governors in the prison service had been reluctant to participate in incident command training. The formalisation of command training has certainly increased in recent years with command doctrines benefiting from proper studies of core command skills (Fallesen, 2000).

Our authors agreed that the experiences gained during training prove their worth at 'show time'. Sarna touched on an important lesson for the trainers: make sure you are teaching the right lesson. Far too often the design and facilitation of exercises is left to junior staff, competent at levels or skills different to those ostensibly being taught. It also pays to be clear, as Salas points out, who or what the subject of the training is. Whereas some will be targeted at individual development, most of it should reflect the fact that the 'teams are the trainees of interest'. Training must be delivered against clear frameworks and objectives; processes must become second nature; there is a need to have certain things drilled/repeated to ensure robustness.

Command training programmes should be based on an up to date analysis of the key skills required for current post holders as well as the commanders of the future. The aviation industry is very advanced in the analysis of non-technical (social and cognitive) skills that can be trained to enhance high level performance in commanders and in more junior staff (see Salas et al, 2001). Dallaire (2000) identifies a need for a more rigorous analysis of military commanders' experiences to extract the lessons to be learned for future commanders in the field. This not only contributes to a training need analysis but if these findings can be presented in an accessible fashion, then they are valuable for continuing professional development as well as initial command training. The underlying thesis of this volume is that there are lessons, which can be usefully transferred across professional domains for command.

Salas et al provided guidance on team training, believing that teams should engage in cross training to gain a complete understanding of their joint roles. This fosters an appreciation of how they can best support their team-mates and achieve successful outcomes in a co-ordinated way. They feel that the optimum method of assessing team performance is by direct observation of the team performing in action, or in 'dynamic manner'. This method should enable an evaluation of the team's performance to be generated against specific behaviours and outcomes. Assessment of competence would ideally, therefore, be integral to the delivery of training and should be done as unobtrusively as possible. Teams must learn continuously; this necessitates appropriate behaviour and attitudes in team members. A willingness to learn and self-evaluate might be held to be an essential characteristic of an aspiring commander or command team member.

Most learning probably takes place outside the classroom setting. Lodge describes graphically his experiences from the flightdeck, contrasting the inhibiting effects of intimidating senior officers with the positive educational influence of the 'benign deities'. A number of contributors described how they had learned by watching more experienced commanders – and the importance of role modelling for command skills development should not be underemphasised.

Competence Assessment

Competences are very specific to roles. There are very few command competences that remain unchanged in the transition from task level, or crew command, to strategic command level. Even the skills surrounding a task which appears the same at any level, such as briefing a group of subordinates, is different according to the level of the commander, the subordinates and the task itself. This must be appreciated when attempting to define roles, and the functional competences that support them.

Assessment of commanders', or command teams', performance is not easy and tends to fall into two distinct categories; the quantitative and the qualitative. Quantitative assessments tend to start at a superficial level observing whether key tasks are performed, appropriate decisions are observed to have been made and all necessary functions performed. This is ideally conducted against a framework of agreed and defined competences. Qualitative assessments are necessarily more subjective and target the soft skills of the commander being assessed. They can necessarily only be conducted by experienced and trained assessors, as the process is centred more on judgement than measurement. Most assessments are based on a framework of quantitative assessment but include a judgement or conclusion based on an element of qualitative assessment of the performance. In recent years there has been a recognition of the need for more rigorous assessment of command competence and many professions have now developed formal standards of competence for incident commanders (see Flin, 1996).

The Command Organisation

Incident Command Systems, Tiers of Command

Currently accepted models portray levels of command that are as much traditional as functional. They date from times when physical distance equated to isolation from information. This is clearly no longer the case, with myriad devices available to supply the commander with images, facts and voice communication from anywhere in the world. Are previously accepted protocols about the senior commander respecting the subordinate commander's clearer awareness of, and better feel for, the situation on the ground now compromised to some extent, or even redundant?

Also, we have seen example after example in this volume of commanders considering it necessary to be near the action, whether for the better understanding of the conditions this permits, or with the purpose of strengthening the morale and resolve of the 'team'. But is this an excuse? Is the natural instinct of senior commanders to feel that they can really do the job on the ground better than their juniors? Or might the reason be one of comfort? The fact is that the task level of operations, at the 'nitty-gritty' end of things, apart from being more dangerous, glamorous, high profile etc., is usually easier to understand and get to grips with. A senior commander who has previously served for a long time at this lower or middle ranking level may well feel more secure and comfortable operating in this sphere.

This does not necessarily question their strategic or tactical management skills, but recognises that commanders who have risen through their organisations have often spent more time, and gained more first hand experience, at levels where responsibility is much lower than that associated with the rank which they currently hold. The questions that arise are: where should the commander ideally exercise command from, and what benefits, real or perceived, are there in the commander's presence at the sharp end of things? Also how ready should she or he be to make an intervention? Commanders who are uncomfortable with the responsibility of their roles demonstrate one or more of the disorders of command decision making; some may hide behind 'being strategic', and others return to their roots, at the sharp end. However, the likelihood is that the best practice is able to be formalised to some degree in a system.

The practical implications of a senior commander choosing to be present on the 'front line' are complex. There is perhaps, a personal need on behalf of the commander to obtain a 'feel' for the incident. Moore, Keeling and Larken felt the need to achieve this degree of empathy with those being led.

The commander who is located at a forward position will, in all probability not be at the optimum point within the communications network, therefore, any communication will have to be relayed. This can impede the flow of information and slows any decision-making that has to be done. In the absence of the decision-makers, others will either take the decisions themselves or situations will remain unresolved. The hierarchical relationships within the UK's 'Gold, Silver, Bronze' protocol (Home Office, 1997) are definitely compromised if the commander from a higher level unnecessarily, or prematurely, intervenes at a lower level. Intervention may be necessary, but there are routes and methods to achieve that. Turning up on the ground and becoming embroiled in the operations is not usually conducive to a successful conclusion to the operation.

Irrespective of the commander's style, location, or predisposition to make tactical interventions, all command decisions have to be acted upon and implemented, otherwise the process is undermined and meaningless. The progress of actions following commands must be tracked to completion. This is part of the necessary information loop that the commander must be at the centre of, to be in a position to fully support the operation in hand.

Communication

Communication is critical to the effective command of fast-developing incidents and all of the contributors have emphasised how problems can hamper the operation. Communication problems take several forms. The simplest breakdown is where actual voice or data communication links fail. If this occurs, personnel have to operate on the basis of understanding the commander's intent and the mission objective. As discussed in Chapter 2, this is one of the defining features of a mature and effective system of command and one of the benefits of the flexibility offered by tight discipline and clear organisational structures. Contrary to some fashionable but ill-informed views expressed by critics of 'militaristic' styles of management, the greatest degrees of latitude can be offered to those who understand systems and objectives and adhere to them in a self-disciplined way. True empowerment can only be offered within a clear framework of rules and unambiguous organisational values. The clearer the framework of command, the more fluid the control network can be.

Therefore, to enable the troops to continue to progress the mission effectively when the commander is absent for any reason, robust communications must be built into a command protocol. It is critically important that command briefings succeed in imparting a deep understanding of the mission objectives and the commander's intent as opposed to a hazy or confused idea of the immediate next steps in the plan. Without this complete understanding, when lines of immediate communication are broken or suspended, the operation will become compromised and freelancing may start to occur. Properly trained and briefed subordinate commanders, in contrast, carry on towards their shared objective unimpeded.

Studies which support this point of view have shown that control and even efficiency advantages can be realised by giving detailed or specific direction to staff who are either less competent or who work with a stable external environment. However, if the task is turbulent or unpredictable, or the workforce comprises more skilled or resourceful people, 'latitude for subordinate improvisation may be vital to organisational effectiveness' (Bigley & Roberts, in press). Nevertheless, it is essential to bear in mind that the same research suggests that careless or lazy delegation disguised as empowerment can lead to serious problems. There is a threat of serious organisational dysfunction when empowered individuals are operating in turbulent environments without adequate means of sufficiently integrating their behaviours, either with each other or their organisation's objectives (Bigley & Roberts, in press).

Commanders and command organisations have long appreciated the need for structure in the uncertain and dynamic environment of command. The more that can be taken out of the realm of the uncertain and made plain, the better. In this category can be placed command structures, terminology, reporting lines, command objectives and familiarity with the team, all of which have been underlined to greater or lesser degrees by the contributors. This conclusion reinforces the need for clear incident command procedures as well as team-focused training under testing conditions.

A common form of communication breakdown occurs when the words that are transmitted or spoken are apparently heard or read by the commander but not able to be assimilated for a variety of reasons. This may be due to the dynamic environment, stress or pure overload of information. It may be due to inexperience in the particular type of situation, rendering the commander unable to identify key parts of information from the whole received. Alternatively, the commander may be exhibiting weaknesses in the categories of encystment or vagabonding and not keeping a full overview. Whichever the cause, research suggests that a disturbingly high proportion of any communication in command environments fails to have the desired impact (Gilchrist, 2000).

Communication as an issue was highlighted by Brunacini, who expressed it as the need for the commander to manage an overall game plan that connects operational action to incident operations. There are at least two contexts; first, the technical or functional context centring on the hardware and procedures, and secondly, the context of communicating the commander's intent.   In the second context, there is a view that understanding and awareness is a group level phenomenon. This states that cognition occurs not only within groups but also between and amongst them (Bigley & Roberts, in press), further strengthening the view of Salas et al that the team is the key entity in dynamic situations.

The challenges presented by modern communication, which facilitate real time management of task level issues by tactical commanders, or of tactical issues by strategic or political level interests has been discussed in some of the preceding chapters. The consensus appears to be that this is an unhelpful development that requires careful and sensitive management and control. A degree of balance must be achieved which must result in the higher command tiers being fully informed but which at the same time discourages the kind of micromanagement that distracts and compromises the commander on the ground from his or her expert management of the job in hand.

Political, Media and Legal Considerations

Political Interest

Command at the higher levels inevitably has a political dimension to it. This might be taken, for the purposes of this discussion, as any consideration affecting the commander's discretion that is extraneous to the technical or immediately practical implications of the task in hand. This applies whether the task happens to be control of a prison riot, a military engagement or a protracted, televised rescue being carried out by firefighters. The definition is further restricted for this brief analysis to embrace only external political dynamics rather than internal or organisational politics, which are active in any organisation.

Politicians at local and national levels have a strategic role in almost every type of organisation whose command methodologies and environments have been considered in this text. They may represent governments controlling their military forces; state, county or city regimes controlling their police forces; or other forms of locally elected body controlling fire brigades. This is right and proper in any democracy. However, it is essential that some form of 'social contract' exists between the politicians and their senior officials about where boundaries of responsibility start and finish. This must recognise the different expertise and responsibilities, which come into play in any given situation, particularly a high profile event that might represent a high risk to any of the stakeholders. It is essential that a politician in such a realm is able and prepared to delegate operational matters entirely to the senior professional. To do otherwise would necessarily distort and impede the operation in progress.

Until relatively recently in historical terms, it was impossible for remotely located stakeholders to influence any level other than the strategic. The state of the art of information and communications technology in general would not have supported an attempt to be in more immediate touch with the detail. Today, however, we have seen how remote stakeholder influences are making themselves felt in the command process. There is clear evidence in the media, as well as from the contributors to this volume, that the concept of separation of responsibility, wherein the politician would be exclusively confined to the strategic, is being eroded.

This problem is far from new. Gordon observed that the British Admiralty experienced the problem early in the last century shortly after the advent of wireless telegraphy (W/T). 'Their Lordships particularly the hyper-meddling Churchill, helped themselves to the facilities of W/T but were loath to acknowledge their share of the operational responsibility'. Also, '.. admirals at sea ... fearful of blighting their careers, put up with the intrusion'. (Gordon, 1996, p. 389).

Strategic commanders must be able to facilitate and support the operation, monitor progress towards objectives, and review outcomes. The majority of the stresses of command are found 'in the hot seat'. This is not, on the whole, located in strategic command positions but quite often at tactical command and very often at task command level. Strategic stakeholders should not, therefore, indulge themselves in a degree of interest that borders on interference, thereby making their subordinates' tasks harder.

This is not to say that an experienced senior commander, based on information and judgement, should not make an intervention if that is judged necessary; the key is in determining when, and to what extent, it may in fact be necessary. This is a command skill that cannot be proceduralised, therefore, it would be wrong to prevent a strategic commander from intervening at whatever level and time required; indeed, some such interventions are part of a 'coaching' approach advocated by experienced commanders. Coyle reminds us that even when an intervention has been made, it can only be in the nature of a broad change of direction and that having exercised command, matters are once again in the hands of the junior commander.

Media Scrutiny

In the military context, the effect of the media enabling politicians and the public to see the immediate environment of the troops and commanders and make their own subjective judgements about them must also cause greater reflection by middle ranking officers in difficult situations. This will not always be bad and may deter a potential maverick in the rare cases where such an attitude exists. However, the concern is that in the wide range of circumstances where it has been seen that only a quick and decisive approach will do, a defensive or passive stance will be taken in preference to one which could attract criticism. Or at least, an already difficult situation for the commander will be made more so.

While industrial commanders and others working on their own premises may be isolated from the intrusion of the media while handling the event, the same protection is not afforded to the emergency services commanders. Press and broadcast reporters normally learn of incidents before emergency services have been fully deployed. They arrive at the scene of the event with extraordinary speed and will usually want to interview the commander in charge of the emergency response rather than a public relations officer or a junior commander. This can be distracting for a busy commander, yet the media requirement for information needs to be met in a professional manner, otherwise they will acquire it in more disruptive ways or from unreliable sources. As emphasised above, this requires having well thought out systems and procedures rehearsed for dealing with the media who arrive on-scene at major incidents. Commanders need to be aware that their decision making will be scrutinised by the on-site media representatives and that the reporters' judgement of the commander's competence may be broadcast while the operation is still running.

The phenomenon of commanders having to deal with public scrutiny in democratic societies is far from new. It is part and parcel of the responsibilities one takes on as a commander in any service or discipline, including in the commercial operations discussed elsewhere in this text. In the guise of the on-scene media it is, without doubt, the source of additional stress for the commander. It is well known that the press are likely to be the commander's friend and admirer in the first hour, critic the rest of the day, and assassin the rest of the week until the event falls off the shelf as old news.

The world media provide quick coverage of breaking news wherever it happens, and that is probably good. This permits rapid analysis by pundits who presume to understand the environment and the operation as well as those directly involved. This may be so, in which case all is well and good;

or it may not be so, in which case destructive and negative forces can come to bear. It can do little to inspire confidence in the leader of an operation on the ground to be conscious that his or her seniors, employers or other stakeholders are second-guessing decisions being taken in real time from the comfort and safety of their office or lounge, but this is increasingly the reality.

The relationship is not always negative, as Larken has described in this volume, but the influence can hardly be ignored and dealing with media is another competence that commanders lower down the hierarchy than in the past now have to have.

Legal Undertones

Some experienced commanders wonder whether a new and perhaps excessive caution is entering the risk assessment process of military and emergency service commanders. There is little doubt that the influence of decisions of courts and tribunals, and, in the UK, the health and safety regulator's (HSE) interventions, have had a significant bearing on this, even though there has been a reluctance on the part of the courts to second guess the decisions of the on-scene commander.

For instance, Mr Justice McNair said in *Bolam v. Friern Hospital Management Committee* [1957] 1 W.L.R. 582.

> A firefighter is not guilty of negligence if he has acted in accordance with practice accepted as proper by a responsible body of firefighting men skilled in that particular art ... merely because there was a body of opinion who would take a contrary view.

A similar judgement has been reported in relation to police commanders, (*Hughes v. National Union of Mineworkers and Others* [1991] ICR 669).

> It will no doubt often happen that in such circumstances critical decisions have to be made with little or no time for considered thought and where many individual officers may be in some kind of physical danger or another. It is not in the public interest that those decisions should generally be the potential target of a negligence claim if rioters do injure an individual officer, since the fear of such a claim would be likely to affect the decisions to the prejudice of the task which the decisions are intended to advance (p. 680).

However there have been at least two earlier legal cases where police commanders have been found culpable and their decision making criticised, see Flin (1996, p. 220).

In British industry there are current government initiatives to strengthen the law on involuntary manslaughter (corporate killing) for cases of injury or death resulting from a company's management failings (see British Safety Council, 2001; Slapper, 1999). To what extent this will influence legal investigations into incident command failings remains to be seen, but it is evident that the degree of accountability being demanded from organisations operating in high-risk domains is increasing.

Future Directions

Criticism of commanders, when it occurs, mainly focuses on decisions that have resulted in loss of life or injury, and, to a lesser extent, property losses. This is right and proper and commanders accept this responsibility. However, it raises the question about whether sufficient structure is attached to the risk assessment skills of command, and to what extent commanders are able to fall back upon reliable and accepted decision making frameworks. Dynamic risk assessment is recognised as a key skill in command and control already, in its many different guises, which are mainly based on various cost benefit algorithms. However it may be considered feasible to develop a 'standard' risk assessment framework, to whatever extent that may be possible. Such a framework would naturally amount to some assessment of the range of risks that may be present in a given situation, the likelihood of each unwanted event occurring, and an estimate of the damage that would result if it did. Although such a device would not be able to be used in the heat of the moment, and therefore would not directly inform the 'dynamic' critical decision, nevertheless, it could offer a framework both for teaching and debriefing. This would lead to clear benefits for the purposes of judicial and internal enquiries, development of risk assessment systems and the sharing of experience and expertise in the use of the systems between different disciplines and services. As with any new approach, it would be vulnerable in its early incarnations from accusations of being simplistic and shallow, however, such steps are essential if risk assessment in emergency situations is to advance from art to science.

Such a process of demystification or transparency would naturally lead to some demand for practitioners to be formally deemed competent in the systems underpinning their roles. Relatively few of the key skills and attributes of senior commanders have been formally catalogued, beyond the general categories discussed earlier. This is not surprising, as there has been little agreement in the matter by various experts and practitioners.

The changing nature of command means that new skills and solutions are required to meet unprecedented challenges. This may be most sharply observed in the military domain (see Keeling's account of protecting the Kurds) as operations other than war begin to dominate the military obligations. McCann and Pigeau's (2000) excellent volume on military command provides several lucid accounts of the new landscape of military operations and the very real difficulties facing commanders and their troops in adapting to these volatile and unfamiliar conditions. Similar changes are present for commanders in other domains with escalation in technological development and civilian crisis situations beginning to take a very different guise from those they may have been trained to anticipate (Rosenthal, Boin & Comfort, in press). Commanders and command organisations will need high level skills and the ability to adapt rapidly to what Davies describes as 'unravelling scenarios'. Crego also emphasises the distinction between fast and 'slow burn' scenarios and commanders need to ensure that their expertise is maintained to enable them to keep apace with emerging and wide ranging threats. Such skills are not in fact peculiar to incident commanders and it is becoming apparent that this type of expertise is also required for the more general business world where risk management for a rapidly changing world requires skills of situation assessment, decision making, team co-ordination and leadership as well as organisational flexibility (Bigley & Roberts, in press).

Conclusion

In conclusion, the main impression left by the accounts of operational commanders, as well as of those who study and teach them, is that little has changed over time in terms of the challenges that commanders face. The ability to generate a clear picture of a confusing situation, to think strategically while simultaneously exercising effective control, and to utilise whatever means of communication is available to best effect, remain important, but these skills have taken on new guises as technology has advanced. However, the need for intelligence, integrity, courage, humility, resilience, enthusiasm and most of the other conventionally accepted traits of the effective leader appear to be as essential in the commander as they ever were.

Discussion throughout this volume has used language that might suggest that a distinction of some kind between the concepts of command, leadership and management is accepted. In fact, although few of the contributors have explicitly referred to the issue, it appears that these distinctions are mainly semantic, rather than a reflection of real underlying differences in basic command skills. However it is clear from the commanders' personal accounts that many of the subtleties of their command expertise are culturally specific. They have all acquired their expertise "on the job". Steeped in the ways, methods and cultures of their organisations, they have adapted to the unique demands of their work environments. Thus the development of their skills and competences has more in common with tacit learning characteristic of apprenticeships, (and increasingly recognised as key to the development of expertise in other professions such as medicine) rather than a formal educational process. There is much to learn from the study of command across disciplines which can improve the training of basic command skills. However, the way these generic skills are applied will still need to be customised for given roles in the culture of a particular organisation.

We hope that this volume will help to raise the level of debate and discussion on the subject of incident command. Our incident commanders' tales have highlighted the wealth of knowledge and expertise that can be gleaned from those who spent their careers 'Sitting in the Hot Seat'.

Note

1.   For details see the RAeS website – www.raes-hfg.com/xsitass.htm

References

Bigley, G. & Roberts, K. (in press) The incident command system: High reliability organising for complex and volatile task environments. *Academy of Management Journal.*
British Safety Council (2001) *Corporate Killing.* London: Author.
Buck, G. (1999) The Role of Cognitive *Complexity in the Management of Critical Situations.* PhD Thesis. University of Westminster.
Cherrie, S. (2000) The human in command. A personal view. In C. McCann & R. Pigeau (Eds.) *The Human in Command. Exploring the Modern Military Experience.* New York: Kluwer/ Plenum.
Crabbe, R. (2000) The nature of command. In C. McCann & R. Pigeau (Eds.) *The Human in Command. Exploring the Modern Military Experience.* New York: Kluwer/ Plenum.
Crichton, M. (in prep) *Training for Emergency Management in the Nuclear Power Industry.* PhD Thesis, University of Aberdeen.
Driskell, J. & Salas, E. (1996) (Eds.) *Stress and Human Performance.* Hillsdale, N.J: Lawrence Erlbaum Associates.
Eid, J. (2000) *Early Predictors of PTSD Symptom Reporting.* PhD thesis, Department of Clinical Psychology, University of Bergen.
Everts, P. (2000) Command and control in stressful situations. In C. McCann & R. Pigeau (Eds.) *The Human in Command. Exploring the Modern Military Experience.* New York: Kluwer/ Plenum.

Fallesen, J. (2000) Developing practical thinking for battle command. In C. McCann & R. Pigeau (Eds.) *The Human in Command. Exploring the Modern Military Experience.* New York: Kluwer/ Plenum.

Fletcher, G., McGeorge, P., Flin, R., Glavin, R. & Moran, N. (in press). The role of non-technical skills in anaesthesia. A review of the literature. *British Journal of Anaesthesia.*

Flin, R. (1996) *Sitting in the Hot Seat. Leaders and Teams for Critical Incident Management.* Chichester: Wiley.

Flin, R. (2001) Selecting the right stuff: Personality and high reliability occupations. In B. Roberts & R. Hogan (Eds.) *Personality Psychology for the Workplace.* Washington: APA Books.

Gilchrist, I. (2000) *An Analysis of the Management of Information on the Fire Service Incident Ground.* PhD Thesis. University of Manchester.

Gordon, A. (1996) *The Rules of the Game – Jutland and the British Naval Command.* London: John Murray.

Howard, A. & Choi, M. (2000) How do you assess a manager's decision making capabilities? The use of situational inventories. *International Journal of Selection and Assessment,* 8, 2, 85-88.

Labbe, S. (2000) Time, tempo and command. In C. McCann & R. Pigeau (Eds.) *The Human in Command. Exploring the Modern Military Experience.* New York: Kluwer/ Plenum.

McCann, C. & Pigeau, R. (2000) (Eds.) *The Human in Command. Exploring the Modern Military Experience.* New York: Kluwer/ Plenum.

Orasanu, J. (1995) Situation awareness: its role in flight crew decision making. In R. Jensen (Ed) *Proceedings of the Eighth Symposium on Aviation Psychology.* Columbus: Ohio State University.

Roberts, B. & Hogan, R. (2001) (Eds.) *Personality Psychology for the Workplace.* Washington: American Psychological Association Books.

Rosenthal, U., Boin, A. & Comfort, L. (in press) (Eds.) *Managing Crises: Threats, Dilemmas, Opportunities.* Springfield, Illinois: C.C. Thomas.

Salas, E., Bowers, C. & Edens, E. (2001) (Eds.) *Improving Teamwork in Organizations. Applications of Resource Management Training.* Mahwah, NJ: Lawrence Erlbaum Associates.

Salas, E. & Klein, G. (2001) (Eds.) *Linking Expertise and Naturalistic Decision Making.* Mahwah, NJ: Lawrence Erlbaum Associates.

Slapper, G. (1999) *Blood in the Bank. Social and Legal Aspects of Death at Work.* Aldershot: Ashgate.

Tissington, P. (2001) *Decision Making by Fireground Incident Commanders.* PhD Thesis, University of Aberdeen.

Woodward, S. (1992) *One Hundred Days. The Memoirs of the Falklands Battle Group Commander.* London: Harper Collins.

Yule, W. (1999) *Post Traumatic Stress Disorders - Concepts and Therapy.* Chichester: Wiley.

Index